Workload Measures

Workload Measures

Authored by
Valerie Jane Gawron

CRC Press
Taylor & Francis Group
Boca Raton London New York

CRC Press is an imprint of the
Taylor & Francis Group, an **informa** business

CRC Press
Taylor & Francis Group
6000 Broken Sound Parkway NW, Suite 300
Boca Raton, FL 33487-2742

© 2019 by Taylor & Francis Group, LLC
CRC Press is an imprint of Taylor & Francis Group, an Informa business

No claim to original U.S. Government works

Printed on acid-free paper

International Standard Book Number-13 978-0-367-00232-9 (Hardback)

This book contains information obtained from authentic and highly regarded sources. Reasonable efforts have been made to publish reliable data and information, but the author and publisher cannot assume responsibility for the validity of all materials or the consequences of their use. The authors and publishers have attempted to trace the copyright holders of all material reproduced in this publication and apologize to copyright holders if permission to publish in this form has not been obtained. If any copyright material has not been acknowledged, please write and let us know so we may rectify in any future reprint.

Except as permitted under U.S. Copyright Law, no part of this book may be reprinted, reproduced, transmitted, or utilized in any form by any electronic, mechanical, or other means, now known or hereafter invented, including photocopying, microfilming, and recording, or in any information storage or retrieval system, without written permission from the publishers.

For permission to photocopy or use material electronically from this work, please access www.copyright.com (http://www.copyright.com/) or contact the Copyright Clearance Center, Inc. (CCC), 222 Rosewood Drive, Danvers, MA 01923, 978-750-8400. CCC is a not-for-profit organization that provides licenses and registration for a variety of users. For organizations that have been granted a photocopy license by the CCC, a separate system of payment has been arranged.

Trademark Notice: Product or corporate names may be trademarks or registered trademarks, and are used only for identification and explanation without intent to infringe.

Visit the Taylor & Francis Web site at
http://www.taylorandfrancis.com

and the CRC Press Web site at
http://www.crcpress.com

To my parents: Jane Elizabeth Gawron 12 June 1926 to 17 March 2002

and Stanley Carl Gawron 17 March 1921 to 9 February 2000.

Contents

List of Figures xi
List of Tables xiii
Preface xv
Acknowledgments xvii
Author xix

1 Introduction 1

2 Human Workload 3
   2.1   Stand-Alone Performance Measures of Workload 5
        2.1.1   Aircrew Workload Assessment System 6
        2.1.2   Control Movements/Unit Time 6
        2.1.3   Glance Duration and Frequency 7
        2.1.4   Load Stress 8
        2.1.5   Observed-Workload Area 9
        2.1.6   Rate of Gain of Information 9
        2.1.7   Relative Condition Efficiency 10
        2.1.8   Speed Stress 10
        2.1.9   Task Difficulty Index 11
        2.1.10 Time Margin 12
   2.2   Secondary Task Performance Measures of Workload 13
        2.2.1   Card-Sorting Secondary Task 15
        2.2.2   Choice RT Secondary Task 16
        2.2.3   Classification Secondary Task 20
        2.2.4   Cross-Adaptive Loading Secondary Task 21
        2.2.5   Detection Secondary Task 22
        2.2.6   Distraction Secondary Task 23
        2.2.7   Driving Secondary Task 24
        2.2.8   Identification/Shadowing Secondary Task 25
        2.2.9   Lexical Decision Secondary Task 27
        2.2.10 Memory Recall Secondary Task 28
        2.2.11 Memory-Scanning Secondary Task 28
        2.2.12 Mental Mathematics Secondary Task 33
        2.2.13 Michon Interval Production Secondary Task 37
        2.2.14 Monitoring Secondary Task 39
        2.2.15 Multiple Task Performance Battery of Secondary Tasks 45
        2.2.16 Occlusion Secondary Task 45
        2.2.17 Problem-Solving Secondary Task 46
        2.2.18 Production/Handwriting Secondary Task 48
        2.2.19 Psychomotor Secondary Task 49

	2.2.20	Randomization Secondary Task 50
	2.2.21	Reading Secondary Task...52
	2.2.22	Simple Reaction-Time Secondary Task 53
	2.2.23	Simulated Flight Secondary Task 55
	2.2.24	Spatial-Transformation Secondary Task..............................56
	2.2.25	Speed-Maintenance Secondary Task 57
	2.2.26	Sternberg Memory Secondary Task......................................57
	2.2.27	Three-Phase Code Transformation Secondary Task 63
	2.2.28	Time-Estimation Secondary Task..63
	2.2.29	Tracking Secondary Task...67
	2.2.30	Workload Scale Secondary Task 71
2.3	Subjective Measures of Workload..71
	2.3.1	Comparison of Subjective Workload Measures 75
		2.3.1.1 Analytical Hierarchy Process...............................75
		2.3.1.2 Magnitude Estimation..78
		2.3.1.3 Pilot Subjective Evaluation...................................79
		2.3.1.4 Subjective Workload Dominance.........................80
	2.3.2	Decision Tree Subjective Workload Measures....................81
		2.3.2.1 Bedford Workload Scale..81
		2.3.2.2 Cooper-Harper Rating Scale.................................84
		2.3.2.3 Honeywell Cooper-Harper Rating Scale.............87
		2.3.2.4 Mission Operability Assessment Technique......88
		2.3.2.5 Modified Cooper-Harper Rating Scale 89
		2.3.2.6 Sequential Judgment Scale....................................93
	2.3.3	Set of Subscales Subjective Workload Measures................95
		2.3.3.1 Assessing the Impact of Automation on Mental Workload (AIM)..95
		2.3.3.2 Crew Status Survey..97
		2.3.3.3 Finegold Workload Rating Scale 100
		2.3.3.4 Flight Workload Questionnaire 102
		2.3.3.5 Hart and Hauser Rating Scale.............................102
		2.3.3.6 Human Robot Interaction Workload Measurement Tool..103
		2.3.3.7 Multi-Descriptor Scale..104
		2.3.3.8 Multidimensional Rating Scale 104
		2.3.3.9 Multiple Resources Questionnaire 106
		2.3.3.10 NASA Bipolar Rating Scale.................................107
		2.3.3.11 NASA Task Load Index 110
		2.3.3.12 Profile of Mood States...133
		2.3.3.13 Subjective Workload Assessment Technique 135
		2.3.3.14 Team Workload Questionnaire 145
		2.3.3.15 Workload/Compensation/Interference/ Technical Effectiveness.. 146
	2.3.4	Single Number Subjective Workload Measures............... 148
		2.3.4.1 Air Traffic Workload Input Technique.............. 148

		2.3.4.2	Continuous Subjective Assessment of Workload 149
		2.3.4.3	Dynamic Workload Scale................................... 149
		2.3.4.4	Equal-Appearing Intervals 150
		2.3.4.5	Hart and Bortolussi Rating Scale...................... 151
		2.3.4.6	Instantaneous Self Assessment (ISA)................ 151
		2.3.4.7	McDonnell Rating Scale 154
		2.3.4.8	Overall Workload Scale..................................... 154
		2.3.4.9	Pilot Objective/Subjective Workload Assessment Technique 157
		2.3.4.10	Utilization.. 159
	2.3.5	Task-Analysis Based Subjective Workload Measures...... 159
		2.3.5.1	Arbeitswissenshaftliches Erhebungsverfahren zur Tatigkeitsanalyze 159
		2.3.5.2	Computerized Rapid Analysis of Workload 160
		2.3.5.3	McCracken-Aldrich Technique 160
		2.3.5.4	Task Analysis Workload.................................... 161
		2.3.5.5	Zachary/Zaklad Cognitive Analysis................ 162
2.4	Simulation of Workload 163
2.5	Dissociation of Workload and Performance 164

List of Acronyms.. 183

Author Index 187

Subject Index 195

List of Figures

Figure 2.1 Guide for selecting a workload measure 4
Figure 2.2 Sternberg memory task data 58
Figure 2.3 Example AHP rating scale 76
Figure 2.4 Pilot subjective evaluation scale
(from Lysaght et al., 1989, p. 107) 79
Figure 2.5 Bedford Workload Scale 82
Figure 2.6 Cooper-Harper Rating Scale 85
Figure 2.7 Honeywell Cooper-Harper Rating Scale
(from Lysaght et al., 1989, p.86) 87
Figure 2.8 Modified Cooper-Harper Rating Scale 90
Figure 2.9 15-point form of the sequential judgment scale
(Pfender et al., 1994, p. 31) 94
Figure 2.10 AIM-s (http://www.eurocontrol.int/humanfactors/
public/standard_page/SHAPE_Questionnaires.html) 96
Figure 2.11 Crew status survey 98
Figure 2.12 Finegold Workload Rating Scale 101
Figure 2.13 Hart and Hauser Rating Scale 103
Figure 2.14 NASA Bipolar Rating Scale 109
Figure 2.15 NASA TLX Rating Sheet 111
Figure 2.16 WCI/TE scale matrix 147
Figure 2.17 Dynamic workload scale 150
Figure 2.18 McDonnell rating scale. (From McDonnell, 1968, p. 7.) 155
Figure 2.19 Dissociations between performance and subjective
measures of workload as predicted by theory
(Adapted from Yeh and Wickens, 1988, p. 115) 165

List of Tables

Table 2.1 References Listed by the Effect on Performance of Primary Tasks Paired with a Secondary Choice RT Task 17

Table 2.2 References Listed by the Effect on Performance of Primary Tasks Paired with a Secondary Detection Task 22

Table 2.3 References Listed by the Effect on Performance of Primary Tasks Paired with a Secondary Identification Task 26

Table 2.4 References Listed by the Effect on Performance of Primary Tasks Paired with a Secondary Memory Task 29

Table 2.5 References Listed by the Effect on Performance of Primary Tasks Paired with a Secondary Mental Math Task 34

Table 2.6 References Listed by the Effect on Performance of Primary Tasks Paired with a Secondary Michon Interval Production Task 38

Table 2.7 References Listed by the Effect on Performance of Primary Tasks Paired with a Secondary Monitoring Task 41

Table 2.8 References Listed by the Effect on Performance of Primary Tasks Paired with a Secondary Occlusion Task 46

Table 2.9 References Listed by the Effect on Performance of Primary Tasks Paired with a Secondary Problem-Solving Task 47

Table 2.10 References Listed by the Effect on Performance of Primary Tasks Paired with a Secondary Psychomotor Task 50

Table 2.11 References Listed by the Effect on Performance of Primary Tasks Paired with a Secondary Randomization Task 51

Table 2.12 References Listed by the Effect on Performance of Primary Tasks Paired with a Secondary Simple RT Task 54

Table 2.13 References Listed by the Effect on Performance of Primary Tasks Paired with a Secondary Sternberg Task 60

Table 2.14 References Listed by the Effect on Performance of Primary Tasks Paired with a Secondary Time Estimation Task 65

Table 2.15 References Listed by the Effect on Performance of Primary Tasks Paired with a Secondary Tracking Task 68

List of Tables

Table 2.16	Comparison of Subjective Measures of Workload	72
Table 2.17	Definitions of AHP Scale Descriptors	77
Table 2.18	Mission Operability Assessment Technique Pilot Workload and Subsystem Technical Effectiveness Rating Scales	89
Table 2.19	Revised Crew Status Survey (Thomas, 2011, p. 13)	99
Table 2.20	Multidimensional Rating Scale	105
Table 2.21	Multiple Resources Questionnaire	106
Table 2.22	NASA Bipolar Rating Scale Descriptions	108
Table 2.23	NASA TLX Rating-Scale Descriptions	112
Table 2.24	SWAT Scales	135
Table 2.25	Team Workload Questionnaire (Sellers et al., 2014, p. 991)	145
Table 2.26	Instantaneous Self Assessment	152
Table 2.27	POSWAT (Mallery and Maresh, 1987, p. 655)	157
Table 2.28	Determinants of Performance and Subjective Measures of Workload (Adapted from Yeh and Wickens, 1988)	166
Table 2.29	Theory of Dissociation	167
Table 2.30	Description of Studies Reviewed	168
Table 2.31	Summary of Research Reporting Both Performance and Subjective Measures of Workload	170
Table 2.32	Points in Which Workload Increased and Performance Improved	179
Table 2.33	Categorization of above Points	180

Preface

This *Human Performance and Situation Awareness Measures* handbook was developed to help researchers and practitioners select measures to be used in the evaluation of human/machine systems. It can also be used to supplement classes at both the undergraduate and graduate courses in ergonomics, experimental psychology, human factors, human performance, measurement, and system test and evaluation. Volume 1 of this handbook begins with an overview of the steps involved in developing a test to measure human performance, workload, and/or situational awareness. This is followed by a definition of human performance and a review of human performance measures. Situational Awareness is similarly treated in subsequent chapters. Finally, workload is defined and measures described in Volume 2.

Acknowledgments

This book began while I was supporting numerous test and evaluation projects of military and commercial transportation systems. Working with engineers, operators, managers, programmers, and scientists showed a need for both educating them on human performance measurement and providing guidance for selecting the best measures for the test. I thank Dr. Dave Meister who provided great encouragement to me to write this book based on his reading of my "measure of the month" article in the Test and Evaluation Technical Group newsletter. He, Dr. Tom Enderwick, and Dr. Dick Pew also provided a thorough review of the first draft of the first edition of this book. For these reviews I am truly grateful. I miss you, Dave.

Author

Valerie Gawron has a BA in Psychology from the State University College at Buffalo, a MA also in Psychology from the State University College at Geneseo, a PhD in Engineering Psychology from the University of Illinois, and a MS in Industrial Engineering and MBA, both from the State University of New York at Buffalo. She completed postdoctoral work in environmental effects on performance at the New Mexico State University in Las Cruces and began work for Calspan directly following. She remained at Calspan for 26 years until it was eventually acquired by General Dynamics and she was made a technology fellow. She is presently a human factors engineer at the MITRE Corporation. Dr. Gawron has provided technical leadership in Research, Development, Test, and Evaluation of small prototype systems through large mass-produced systems, managed million dollar system development programs, led the design of information systems to support war fighters and intelligence personnel, fielded computer-aided engineering tools to government agencies and industry, tested state-of-the-art displays including Helmet Mounted Displays, Night Vision Goggles, and Synthetic Vision Displays in military and commercial aircraft, evaluated security systems for airports and United States Embassies, conducted research in both system and human performance optimization, applied the full range of evaluation tools from digital models through human-in-the-loop simulation to field operational tests for military, intelligence, and commercial systems, directed accident reenactments, consulted on driver distraction, accident investigation, and drug effects on operator performance, and written over 425 publications including the *Human Performance, Workload, and Situation Awareness Measures Handbook* (second edition) and *2001 Hearts: The Jane Gawron Story*. Both are being used internationally in graduate classes, the former in human factors and the latter in patient safety.

Dr. Gawron has served on the Air Force Scientific Advisory Board, the Army Science Board, the Naval Research Advisory Committee, and the National Research Council. She gives workshops on a wide range of topics to very diverse audiences from parachute testing given as part of the Sally Ride Science Festival for girls ages 8 to 14 to training applications of simulation to managers and engineers. She has worked programs for the United States Air Force, Army, Navy, Marines, NASA, the Departments of State and Justice, the Federal Aviation Administration, the Transportation Security Administration, the National Transportation Safety Board, the National Traffic Safety Administration, as well as for commercial customers. Some of this work has been international and Dr. Gawron has been to 195 countries. Dr. Gawron is an associate fellow of the American Institute of Aeronautics and Astronautics, a fellow of the Human Factors and Ergonomics Society, and a fellow of the International Ergonomics Association.

1
Introduction

Human factors specialists, including ergonomists, industrial engineers, engineering psychologists, human factors engineers, and many others, continually seek better (more efficient and effective) ways to characterize and measure the human element as part of the system so we can build trains, planes, and automobiles, process control stations, and others with superior human/system interfaces. Yet the human factors specialist is often frustrated by the lack of readily accessible information on human performance, workload, and Situational Awareness (SA) measures. To fill that void, this book was written to guide the reader through the critical process of selecting the appropriate measures of human performance, workload, and SA for objective evaluations.

Chapter 2 describes measures of workload. Each measure is described, along with its strengths and limitations, data requirements, threshold values, and sources of further information. To make this desk reference easier to use, extensive author and subjective indices are provided.

2
Human Workload

Workload has been defined as a set of task demands, as effort, and as activity or accomplishment (Gartner and Murphy, 1979). The task demands (task load) are the goals to be achieved, the time permitted to perform the task, and the performance level to which the task is to be completed. The factors affecting the effort expended are the information and equipment provided, the task environment, the participant's skills and experience, the strategies adopted, and the emotional response to the situation. These definitions provide a testable link between task load and workload. For example (paraphrased from an example given by Azad Madni, vice president, Perceptronics, Woodland Hills, CA, on the results of an Army study), the workload of a helicopter pilot in maintaining a constant hover may be 70 on a scale of 0 to 100. Given the task of maintaining a constant hover and targeting a tank, workload may also be 70. The discrepancy results from the pilot self-imposing a strict performance requirement on hover-only (no horizontal or vertical movement) but relaxing the performance requirement on hover (a few feet movement) when targeting the tank to keep workload within a manageable level. These definitions enable task load and workload to be explainable in real work situations. These definitions also enable a logically correct analysis of task load and workload.

Realistically, workload can *never* exceed 100% (a person cannot do the impossible). Any theories or reported results that allow workload to exceed 100% are not realistic. However, as defined, task load may exceed 100%. An example is measuring "time required/time available." By the proposed definition, this task load measurement may exceed 100% if the performance requirements are set too high (thereby increasing the time required) or the time available is set too low. In summary, workload cannot exceed 100% even if task load does exceed 100%.

Workload has been measured as stand-alone performance (see Section 2.1) or secondary task performance (see Section 2.2) or as subjective measures (see Section 2.3) or in digital simulation (see Section 2.4). Physiological measures of workload have also been identified but are not discussed here. An excellent reference on physiological measures of workload is presented in Caldwell et al. (1994). Dissociation between workload and performance is discussed in Section 2.5.

Guidelines for selecting the appropriate workload measure are given in Wierwille et al. (1979) and O'Donnell and Eggemeier (1986). For mental workload see Moray (1982). Wierwille and Eggemeier (1993) listed four aspects of measures that were critical: diagnosticity, global sensitivity, transferability,

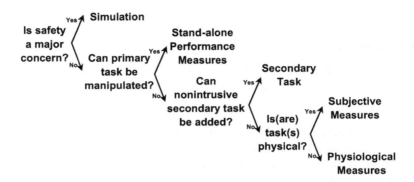

FIGURE 2.1
Guide for selecting a workload measure.

and implementation requirements. A general guide is presented in Figure 2.1. Note that the bottom branch can be repeated for simulation.

A similar categorization of workload measures is presented in Stanton et al. (2005). Their categories are: primary and secondary task performance measures, physiological measures, and subjective-rating measures. Further, Funke et al. (2012) proposed a theory of team workload as well as subjective, performance, physiological, and strategy shift measures. Sharples and Megaw (2015) provided the following classification of workload measures:

1. Analytic techniques: comparative analysis, mathematical models, expert opinion, task analytic methods, and simulation models
2. Empirical techniques: primary task performance, secondary task performance
3. Psychophysiological techniques: cardiac activity, brain activity, electrodermal activity, eye function, body fluid analysis, muscle and movement analysis
4. Subjective/operator opinion techniques: single-dimensional scales, multidimensional scales, relative judgments, instantaneous judgments, interviews, and observations.

Finally, Matthews and Reinerman-Jones (2017) published a book on workload assessment.

Sources

Caldwell, J.A., Wilson, G.F., and Cetinguc, M. *Psychophysiological Assessment Methods (AGARD-AR-324)*. Neuilly-Sur-Seine, France: Advisory Group For Aerospace Research and Development, May 1994.

Funke, G.J., Knott, B.A., Salas, E., Pavlas, D., and Strang, A.J. Conceptualization and measurement of team workload: A critical need. *Human Factors* 54(1): 36–51, 2012.

Gartner, W.B., and Murphy, M.R. Concepts of workload. In B.O. Hartman and R.E. McKenzie (Eds.) *Survey of Methods to Assess Workload (AGARD-AG-246)*. Neuilly-Sur-Seine, France: Advisory Group for Aerospace Research and Development, 1979.

Matthews, G., and Reinerman-Jones, L.E. *Workload Assessment: How to Diagnose Workload Issues and Enhance Performance*. Santa Monica, California: Human Factors and Ergonomics Society, 2017.

Moray, N. Subjective mental workload. *Human Factors* 24(1): 25–40, 1982.

O'Donnell, R.D., and Eggemeier, F.T. Workload assessment methodology. In K.R. Boff, L. Kaufman, and J.P. Thomas (Eds.) *Handbook of Perception and Human Performance*. New York, NY: John Wiley and Sons, 1986.

Sharples, S., and Megaw, T. Definition and measurement of human workload. In J.R. Wilson and S. Sharples (Eds.) *Evaluation of Human Work* (pp. 515–548). Boca Raton: CRC Press, 2015.

Stanton, N.A., Salmon, P.M., Walker, G.H., Barber, C., and Jenkins, D.P. *Human Factors Methods: A Practical Guide for Engineering and Design*. Aldershot, United Kingdom: Gower ebook, December 2005.

Wierwille, W.W., and Eggemeier, F.T. Recommendations for mental workload measurement in a test and evaluation environment. *Human Factors* 35(2): 263–281, 1993.

Wierwille, W.W., Williges, R.C., and Schiflett, S.G. Aircrew workload assessment techniques. In B.O. Hartman and R.E. McKenzie (Eds.) *Survey of Methods to Assess Workload (AGARD-AG-246)*. Neuilly-Sur-Seine, France: Advisory Group for Aerospace Research and Development, 1979.

2.1 Stand-Alone Performance Measures of Workload

Performance has been used to measure workload. These measures assume that, as workload increases, the additional processing requirements will degrade performance. O'Donnell and Eggemeier (1986) identified four problems associated with using performance as a measure of workload: (1) underload may enhance performance, (2) overload may result in a floor effect, (3) confounding effects of information-processing strategy, training, or experience, and (4) measures are task specific and cannot be generalized to other tasks. Meshkati et al. (1990) stated that multiple task measures are required when the task is complex or multidimensional. In addition, task measures may be intrusive and may be influenced by factors other than workload, for example, motivation and learning.

Stand-alone measures of performance include aircrew workload assessment (Section 2.1.1), control movements/unit time (Section 2.1.2), glance duration and frequency (Section 2.1.3), load stress (Section 2.1.4), observational workload area (Section 2.1.5), rate of gain of information (Section 2.1.6), relative condition efficiency (Section 2.1.7), speed stress (Section 2.1.8), Task Difficulty Index (Section 2.1.9), and time margin (Section 2.1.10).

Sources

Meshkati, N., Hancock, P.A., and Rahimi, M. Techniques in mental workload assessment. In J.R. Wilson and E.N. Corlett (Eds.) *Evaluation of Human Work. A Practical Ergonomics Methodology* (pp. 605–627). New York: Taylor & Francis Group, 1990.

O'Donnell, R.D., and Eggemeier, F.T. Workload assessment methodology. In K.R. Boff, L. Kaufman, and J.P. Thomas (Eds.) *Handbook of Perception and Human Performance* (pp. 42-1–42-49). New York: Wiley and Sons, 1986.

2.1.1 Aircrew Workload Assessment System

General description – The Aircrew Workload Assessment System (AWAS) is a timeline analysis software developed by British Aerospace to predict workload. AWAS requires three inputs: (1) second-by-second descriptions of pilot tasks during flight, (2) demands on each of Wickens' multiple resource theory processing channels, and (3) effects of simultaneous demand on a single channel (Davies et al., 1995).

Strengths and limitations – Davies et al. (1995) reported a correlation of +0.904 between AWAS workload prediction and errors in a secondary auditory discrimination task. The participants were two experienced pilots flying a Sea Warrior Simulator.

Data requirements – Second-by-second timeline of pilot tasks, demands on each information processing channel, and effect of simultaneous demand.

Thresholds – Not stated.

Source

Davies, A.K., Tomoszek, A., Hicks, M.R., and White, J. AWAS (Aircrew Workload Assessment System): Issues of theory, implementation, and validation. In R. Fuller, N. Johnston, and N. McDonald (Eds.) *Human Factors in Aviation Operations*. Proceedings of the 21st Conference of the European Association for Aviation Psychology (EAAP), vol. 3, Chapter 48, 1995.

2.1.2 Control Movements/Unit Time

General description – Control movements/unit time is the number of control inputs made summed over each control used by one operator divided by the amount of time over which the measurements were made.

Strengths and limitations – Wierwille and Connor (1983) stated that this measure was sensitive to workload. Their specific measure was the average count per second of inputs into the flight controls (ailerons, elevator, and rudder) in a moving-base flight simulator. The workload manipulation was pitch stability, wind-gust disturbance, and crosswind direction and velocity.

Porterfield (1997), using a similar approach, evaluated the use of the duration of time that an en route Air Traffic Controller was engaged in ground-to-air communications as a measure of workload. He reported a significant correlation (+0.88) between the duration and the Air Traffic Workload Input Technique (ATWIT), a workload rating based on the Pilot Objective/ Subjective Workload Assessment Technique (POSWAT).

Zeitlin (1995) developed a driver workload index based on brake actuations per minute plus the log of vehicle speed. This index was sensitive to differences in type of roadway (rural, city, expressway).

Data requirements – Control movements must be well defined.

Thresholds – Not stated.

Sources

Porterfield, D.H. Evaluating controller communication time as a measure of workload. *The International Journal of Aviation Psychology* 7(2): 171–182, 1997.

Wierwille, W.W., and Connor, S.A. Evaluation of 20 workload measures using a psychomotor task in a moving-base aircraft simulator. *Human Factors* 25(1): 1–16, 1983.

Zeitlin, L.R. Estimates of driver mental workload: A long term field trial of two subsidiary tasks. *Human Factors* 37(3): 611–621, 1995.

2.1.3 Glance Duration and Frequency

General description – The duration and frequency of glances to visual displays have been used as measures of visual workload. The longer the durations and/or the greater the frequency of glances, the higher the visual workload.

Strengths and limitations – Fairclough et al. (1993) used glance duration to calculate the percentage of time that drivers looked at navigation information (a paper map versus a Liquid Crystal Display (LCD) text display), roadway ahead, rear-view mirror, dashboard, left-wing mirror, right-wing mirror, left window, and right window. Data were collected in an instrumented vehicle driven on British roads. The authors concluded that this "measure proved sensitive enough to (a) differentiate between the paper map and the LCD/ text display and (b) detect associated changes with regard to other areas of the visual scene" (p. 248). The authors warned, however, that reduction in glance durations might reflect the drivers' strategy to cope with the amount and legibility of the paper map.

These authors also used glance duration and frequency to compare two in-vehicle route guidance systems. The data were collected from 23 participants driving an instrumented vehicle in Germany. The data indicated "as glance frequency to the navigation display increases, the number of glances to the dashboard, rear-view mirror and the left-wing mirror all show a significant decrease" (p. 251). Based on these results, the authors concluded that "glance duration appears to be more sensitive to the difficulty of information update. Glance frequency represents the amount of 'visual checking behavior'" (p. 251).

Wierwille (1993) concluded from a review of driver visual behavior that such behavior is "relatively consistent" (p. 278).

Data requirements – Record participant's eye position.

Threshold – 0 to infinity.

Sources

Fairclough, S.H., Ashby, M.C., and Parkes, A.M. In-vehicle displays, visual workload and visibility evaluation. In A.G. Gale, I.D. Brown, C.M. Haslegrave, H.W. Kruysse, and S.P. Taylor (Eds). *Vision in Vehicles – IV* (pp. 245–254). Amsterdam: North-Holland, 1993.

Wierwille, W.W. An initial model of visual sampling of in-car displays and controls. In A.G. Gale, I.D. Brown, C.M. Haslegrave, H.W. Kruysse, and S.P. Taylor (Eds.) *Vision in Vehicles – IV* (pp. 271–280). Amsterdam: North-Holland, 1993.

2.1.4 Load Stress

General description – Load stress is the strain produced by increasing the number of signal sources that must be monitored during a task (Chiles and Alluisi, 1979).

Strengths and limitations – Load stress affects the number of errors made in task performance. Increasing load stress to measure operator workload may be difficult in nonlaboratory settings.

Data requirements – The signal sources must be unambiguously defined.

Thresholds – Not stated.

Source

Chiles, W.D., and Alluisi, E.A. On the specification of operator or occupational workload with performance-measurement methods. *Human Factors* 21(5): 515–528, 1979.

2.1.5 Observed-Workload Area

General description – Laudeman and Palmer (1995) developed the Observed-Workload Area to measure workload in aircraft cockpits. The measure is not based on theory but rather on a logical connection between workload and task constraints. In their words:

"An objectively defined window of opportunity exists for each task in the cockpit. The observed workload of a task in the cockpit can be operationalized as a rating of maximum task importance supplied by a domain expert. Task importance increases during the task window of opportunity as a linear function of task importance versus time. When task windows of opportunity overlap, resulting in an overlap of task functions, the task functions can be combined in an additive manner to produce a composite function that includes the observed workload effects of two or more task functions. We called these composites of two or more task functions observed-workload functions. The dependent measure that we proposed to extract from the observed-workload function was the area under it that we called observed-workload area" (pp. 188–190).

Strengths and limitations – Laudeman and Palmer (1995) reported a significant correlation between the first officers' workload-management ratings and the observed-workload area. This correlation was based on 18 two-person aircrews flying a high-fidelity aircraft simulator. Small observed-workload areas were associated with high-workload management ratings. Higher error rate crews had higher observed-workload areas. The technique requires an expert to provide task importance ratings. It also requires well-defined beginnings and ends to tasks.

Thresholds – Not stated.

Source

Laudeman, I.V., and Palmer, E.A. Quantitative measurement of observed workload in the analysis of aircrew performance. *International Journal of Aviation Psychology* 5(2): 187–197, 1995.

2.1.6 Rate of Gain of Information

General description – This measure is based on Hick's Law, which states that reaction time (RT) is a linear function of the amount of information transmitted, (H_t): $RT = a + B\ (H_t)$ (Chiles and Alluisi, 1979).

Strengths and limitations – Hick's Law has been verified in a wide range of conditions. However, it is limited to only discrete tasks and, unless the task is part of the normal procedures, may be intrusive, especially in nonlaboratory settings.

Data requirements – Rate of gain of information is estimated from RT. Time is typically collected with either mechanical stop watches or software clocks. The first type of clock requires frequent (for example, prior to every trial) calibration; software clocks require a stable and constant source of power.

Thresholds – Not stated.

Source

Chiles, W.D., and Alluisi, E.A. On the specification of operator or occupational workload with performance-measurement methods. *Human Factors* 21(5): 515–528, 1979.

2.1.7 Relative Condition Efficiency

General description – Paas and van Merrienboer (1993) combined ratings of workload with task performance measures to calculate relative condition efficiency. Ratings varied from 1 (very, very low mental effort) to 9 (very, very high mental effort). Performance was measured as the percent of correct answers to test questions. Relative condition efficiency was calculated "as the perpendicular distance to the line that is assumed to represent an efficiency of zero" (p. 737).

Strengths and limitations – Efficiency scores were significantly different between work conditions.

Data requirements – Not stated.

Thresholds – Not stated.

Source

Paas, F.G.W.C., and van Merrienboer, J.J.G. The efficiency of instructional conditions: An approach to combine mental effort and performance measures. *Human Factors* 35(4): 737–743, 1993.

2.1.8 Speed Stress

General description – Speed stress is stress produced by increasing the rate of signal presentation from one or more signal sources.

Strengths and limitations – Speed stress affects the number of errors made as well as the time to complete tasks (Conrad, 1956; Knowles et al., 1953). It may be difficult to impose speed stress on nonlaboratory tasks.

Data requirements – The task must include discrete signals whose presentation rate can be manipulated.
Thresholds – Not stated.

Sources

Conrad, R. The timing of signals in skill. *Journal of Experimental Psychology* 51: 365–370, 1956.

Knowles, W.B., Garvey, W.D., and Newlin, E.P. The effect of speed and load on display-control relationships. *Journal of Experimental Psychology* 46: 65–75, 1953.

2.1.9 Task Difficulty Index

General description – The Task Difficulty Index was developed by Wickens and Yeh (1985) to categorize the workload associated with typical laboratory tasks. The index has four dimensions:

1. Familiarity of stimuli
    0 = letters
    1 = spatial dot patterns, tracking cursor
2. Number of concurrent tasks
    0 = single
    1 = dual
3. Task difficulty
    0 = memory set size 2
    1 = set size 4, second-order tracking, delayed recall
4. Resource competition
    0 = no competition
    1 = competition for either modality of stimulus (visual, auditory) or central processing (spatial, verbal) (Gopher and Braune, 1984).

The Task Difficulty Index is the sum of the scores on each of the four dimensions listed above.

Strengths and limitations – Gopher and Braune (1984) reported a significant positive correlation (+0.93) between Task Difficulty Index and subjective measures of workload. Their data were based on responses from 55 male participants performing 21 tasks including the Sternberg, hidden pattern, card rotation tracking, maze tracing, delayed digit recall, and dichotic listening.

Data requirements – This method requires the user to describe the tasks to be performed on the four dimensions given above.
Thresholds – Values vary between 0 and 4.

Sources

Gopher, D., and Braune, R. On the psychophysics of workload: Why bother with subjective measures? *Human Factors* 26(5): 519–532, 1984.

Wickens, C.D., and Yeh, Y. POCs and performance decrements: A reply to Kantowitz and Weldon. *Human Factors* 27: 549–554, 1985.

2.1.10 Time Margin

General description – After a review of current in-flight workload measures, Gawron et al. (1989) identified five major deficiencies: (1) the subjective ratings showed wide individual differences well beyond those that could be attributed to experience and ability differences, (2) most of the measures were not comprehensive and assessed only a single dimension of workload, (3) many workload measures were intrusive in terms of requiring task responses or subjective ratings or the use of electrodes, (4) some measures were confusing to participants in high stress, for example, the meanings of ratings would be forgotten in high-workload environments, so lower than actual values would be given by the pilot, and (5) participants would misperceive the number of tasks to be performed and provide an erroneously low measure of workload. Gawron then returned to the purpose of workload measure: to identify potentially dangerous situations. Poor designs, inadequate procedures, poor training, or the proximity to catastrophic conditions could induce such situations. The most objective measure of danger in a situation is time until the aircraft is destroyed if control action is not taken. These times include: time until impact, time until the aircraft is overstressed and breaks apart, and time until the fuel is depleted.

Strengths and limitations – The time margin measure of workload is quantitative, objective, directly related to performance, and can be tailored to any mission. For example, time until a surface-to-air missile destroys the aircraft is a good measure in air-to-ground penetration missions. In addition, the time margin can be easily computed from measures of aircraft performance. Finally, these times can be calculated over intervals of any length to provide interval-by-interval workload comparisons.

Data requirements – This method is useful whenever aircraft performance data are available.

Thresholds – Minimum is 0 and the maximum is infinity.

Source

Gawron, V.J., Schiflett, S.G., and Miller, J.C. Measures of in-flight workload. In R.S. Jensen (Ed.) *Aviation Psychology* (pp. 240–287). London: Gower, 1989.

2.2 Secondary Task Performance Measures of Workload

One of the techniques used most widely to measure workload is the secondary task. This technique requires an operator to perform the primary task within that task's specified requirements and to use any spare attention or capacity to perform a secondary task. The decrement in performance of the secondary task is operationally defined as a measure of workload.

Advantages. The secondary-task technique has several advantages. First, it may provide a sensitive measure of operator capacity and may distinguish among alternative equipment configurations that are indistinguishable by single-task performance (Slocum et al., 1971). Second, it may provide a sensitive index of task impairment due to stress. Third, it may provide a common metric for comparisons of different tasks.

Disadvantages. The secondary-task technique may have one major disadvantage: intrusion on the performance of the primary task (Williges and Wierwille, 1979). Vidulich (1989a), however, concluded from two experiments that secondary tasks that do not intrude on primary task performance are insensitive to primary-task difficulty. Vidulich (1989b) argued that added task sensitivity is directly linked to intrusiveness.

Damos (1993) analyzed the results of 14 studies in which single- and dual-task performances were evaluated. She concluded that: "the effect sizes associated with both single- and multiple-task measures were both statistically different from 0.0, with the effect size for the multiple-task increased statistically greater than that of the corresponding single task measures. However, the corresponding predictive validities were low" (p. 615). Poulton (1965) pointed out that comparing the results of performance tests that vary in sensitivity may be difficult. To address this concern, Colle et al. (1988) developed the method of double trade-off curves to equate performance levels on different secondary tasks. In this method, "two different secondary tasks are each paired with the same primary tasks. A trade-off curve is obtained for each secondary task paired with the primary task" (p. 646).

Another potential disadvantage is that participants may use different strategies when performing secondary tasks. For example, Schneider and Detweiler (1988) identified seven compensatory activities that are associated with dual-task performance: "(1) shedding and delaying tasks and preloading buffers, (2) letting go of high-workload strategies, (3) utilizing noncompeting resources, (4) multiplexing over time, (5) shortening transmissions, (6) converting interference from concurrent transmissions, and (7) chunking of transmissions" (p. 539).

Wetherell (1981) investigated seven secondary tasks (addition, verbal reasoning, attention, short-term memory, random digit generation, memory search, and white noise) with a primary driving task and concluded that none were "outstanding as a measure of mental workload." However, there was a significant gender difference with degradations in the primary task occurring for female drivers. Further, Ogdon et al. (1979) concluded from a survey of the literature that there was not a single best secondary task for measuring workload. Finally, Rolfe (1971) stated: "The final word, however, must be that the secondary task is no substitute for competent and comprehensive measurement of primary task performance" (p. 146).

Recommendations. To help researchers select secondary task measures of workload, Knowles (1963) developed a comprehensive set of criteria for selecting a secondary task: (1) noninterference with the primary task, (2) ease of learning, (3) self-pacing, (4) continuous scoring, (5) compatibility with the primary task, (6) sensitivity, and (7) representativeness. In a similar vein, Fisk et al. (1983) developed three criteria which they then tested in an experiment. The criteria were: (1) the secondary task must use the same resources as the primary task, (2) single and dual-task performance must be maintained, and (3) the secondary task must require "controlled or effortful processing" (p. 230). However, Liu and Wickens (1987) reported that tasks using the same resources had increased workload more than tasks that did not.

Brown (1978) recommended that "the dual task method should be used for the study of individual difference in processing resources available to handle work-load" (p. 224). Meshkati et al. (1990) recommend not using secondary tasks and subjective measures in the same experiment since operators may include secondary task performance as part of their subjective workload rating.

Sources

Brown, I.D. Dual task methods of assessing work-load. *Ergonomics* 21(3): 221–224, 1978.

Colle, H., Amell, J.R., Ewry, M.E., and Jenkins, M.L. Capacity equivalence curves: A double trade-off curve method for equating task performance. *Human Factors* 30(5): 645–656, 1988.

Damos, D. Using meta-analysis to compare the predictive validity of single- and multiple-task measures of flight performance. *Human Factors* 35(4): 615–628, 1993.

Fisk, A.D., Derrick, W.L., and Schneider, W. The assessment of workload: Dual task methodology. Proceedings of the Human Factors Society 27th Annual Meeting, 229–233, 1983.

Knowles, W.B. Operator loading tasks. *Human Factors* 5: 151–161, 1963.
Liu, Y., and Wickens, C.D. The effect of processing code, response modality and task difficulty on dual task performance and subjective workload in a manual system. Proceedings of the Human Factors Society 31st Annual Meeting, 847–851, 1987.
Meshkati, N., Hancock, P.A., and Rahimi, M. Techniques in mental workload assessment. In J.R. Wilson and E.N. Corlett (Eds.) *Evaluation of a Human Work. A Practical Ergonomics Methodology* (pp. 605–627). New York: Taylor & Francis Group, 1990.
Ogdon, G.D., Levine, J.M., and Eisner, E.J. Measurement of workload by secondary tasks. *Human Factors* 21(5): 529–548, 1979.
Poulton, E.C. On increasing the sensitivity of measures of performance. *Ergonomics* 8(1): 69–76, 1965.
Rolfe, J.M. The secondary task as a measure of mental load. In W.T. Singleton, J.G. Fox, and D. Whitfield (Eds.) *Measurement of Man at Work* (pp. 135–148). London: Taylor & Francis Group, 1971.
Schneider, W., and Detweiler, M. The role of practice in dual-task performance: Toward workload modeling in a connectionist/control architecture. *Human Factors* 30(5): 539–566, 1988.
Slocum, G.K., Williges, B.H., and Roscoe, S.N. Meaningful shape coding for aircraft switch knobs. *Aviation Research Monographs* 1(3): 27–40, 1971.
Vidulich, M.A. Objective measures of workload: Should a secondary task be secondary? Proceedings of the Fifth International Symposium on Aviation Psychology, 802–807, 1989a.
Vidulich, M.A. Performance-based workload assessment: Allocation strategy and added task sensitivity. Proceedings of the Third Annual Workshop on Space Operations, Automation, and Robotics (SOAR'89), 329–335, 1989b.
Wetherell, A. The efficacy of some auditory-vocal subsidiary tasks as measures of the mental load on male and female drivers. *Ergonomics* 24(3): 197–214, 1981.
Williges, R.C., and Wierwille, W.W. Behavioral measures of aircrew mental workload. *Human Factors* 21: 549–574, 1979.

2.2.1 Card-Sorting Secondary Task

General description – "The subject must sort playing cards by number, color, and/or suit" (Lysaght et al., 1989, p. 234).

Strengths and limitations – "Depending upon the requirements of the card sorting rule, the task can impose demands on perceptual and cognitive processes" (Lysaght et al., 1989, p. 234). Lysaght et al. (1989) state that dual-task pairing of a primary memory task with a secondary card-sorting task resulted in a decrement in performance in both tasks. Their statement is based on two experiments by Murdock (1965). Although used as a primary task, Courtney and Shou (1985) concluded that card sorting was a "rapid and simple means of estimating relative visual-lobe size" (p. 1319).

Data requirements – The experimenter must be able to record the number of cards sorted and the number of incorrect responses.

Thresholds – Not stated.

Sources

Courtney, A.J., and Shou, C.H. Simple measures of visual-lobe size and search performance. *Ergonomics* 28(9): 1319–1331, 1985.

Lysaght, R.J., Hill, S.G., Dick, A.O., Plamondon, B.D., Linton, P.M., Wierwille, W.W., Zaklad, A.L., Bittner, A.C., and Wherry, R.J. Operator workload: Comprehensive review and evaluation of operator workload methodologies (Technical Report 851). Alexandria, VA: Army Research Institute for the Behavioral and Social Sciences, June 1989.

Murdock, B.B. Effects of a subsidiary task on short-term memory. *British Journal of Psychology* 56: 413–419, 1965.

2.2.2 Choice RT Secondary Task

General description – "The subject is presented with more than one stimulus and must generate a different response for each one" (Lysaght et al., 1989, p. 232).

Strengths and limitations – "Visual or auditory stimuli may be employed and the response mode is usually manual. It is theorized that choice RT imposes both central processing and response selection demands" (Lysaght et al., 1989, p. 232).

Based on 19 studies that included a choice RT secondary task, Lysaght et al. (1989) stated that, in dual-task pairings: performance of choice RT, problem solving, and flight simulation primary tasks remained stable; performance of tracking, choice RT, memory, monitoring, driving, and lexical decision primary tasks degraded; and tracking performance improved. Performance of the secondary task remained stable with tracking and driving primary tasks; and degraded with tracking, choice RT, memory, monitoring, problem solving, flight simulation, driving, and lexical decision primary tasks (see Table 2.1).

Hicks and Wierwille (1979) manipulated workload by increasing wind gust in a driving simulator. They reported that a secondary RT task was not as sensitive to wind gust as were steering reversals, yaw deviation, subjective opinion rating scales, and lateral deviations. Johnson and Haygood (1984) varied the difficulty of a primary simulated driving task by varying the road width. The secondary task was a visual choice RT task. Tracking score was highest when the difficulty of the primary task was adapted as a function of primary task performance. It was lowest when the difficulty was fixed.

Klapp et al. (1987) asked participants to perform a visual, zero-order, pursuit tracking task with the right hand while performing a two-choice, auditory reaction task with the left hand. In the dual-task condition, the tracking task was associated with hesitations lasting 333 ms or longer. Degradations in the tracking task were associated with enhancements of the RT task. Gawron (1982) reported longer RTs and lower percent correct scores when a four-choice RT task was performed simultaneously then sequentially.

Human Workload

TABLE 2.1
References Listed by the Effect on Performance of Primary Tasks Paired with a Secondary Choice RT Task

Type	Primary Task Stable	Primary Task Degraded	Primary Task Enhanced	Secondary Task Stable	Secondary Task Degraded	Secondary Task Enhanced
Choice RT	Becker (1976) Ellis (1973)	Detweile and Lundy (1995)[a] Gawron (1982)[a] Schvaneveldt (1969)		Hicks and Wierwille (1979)	Becker (1976) Detweiler and Lundy (1995)[a] Ellis (1973) Gawron (1982)[a] Schvaneveldt (1969)	
Driving	Kantowitz (1995)	Allen et al. (1976) Brown et al. (1969) Glaser and Glaser (2015)		Allen et al. (1976) Drory (1985)[a]	Brown et al. (1969) Glaser and Glaser (2015)	
Flight Simulation	Bortolussi et al. (1989) Bortolussi et al. (1986) Kantowitz et al. (1983)[a] Kantowitz et al. (1984)[a] Kantowitz et al. (1987)				Bortolussi et al. (1989) Bortolussi et al. (1986)	
Lexical Decision		Becker (1976)			Becker (1976)	
Memory		Logan (1970)			Logan (1970)	
Monitoring		Smith (1969)			Krol (1971) Smith (1969)	
Problem Solving	Fisher (1975a) Fisher (1975b)				Fisher (1975a) Fisher (1975b)	
Tracking		Benson et al. (1965) Loeb and Jones (1978) Giroud et al. (1984) Israel et al. (1980a,b) Klapp et al. (1980a,b) Israel et al. (1984) Wempe and Baty (1968) Klapp et al. (1987)[a]		Loeb and Jones (1978)	Benson et al. (1965) Damos (1978) Giroud et al. (1984) Israel et al. (1980a,b) Klapp et al. (1984)	Klapp et al. (1987)[a]

Source: From Lysaght et al. (1985) p. 246.
[a] Not included in Lysaght et al. (1989).

Data requirements – The experimenter must be able to record and calculate: mean RT for correct responses, mean (median) RT for incorrect responses, number of correct responses, and number of incorrect responses.
Thresholds – Not stated.

Sources

Allen, R.W., Jex, H.R., McRuer, D.T., and DiMarco, R.J. Alcohol effects on driving behavior and performance in a car simulator. *IEEE Transactions on Systems Man and Cybernetics* SMC-5: 485–505, 1976.

Becker, C.A. Allocation of attention during visual word recognition. *Journal of Experimental Psychology: Human Perception and Performance* 2: 556–566, 1976.

Benson, A.J., Huddleston, J.H.F., and Rolfe, J.M. A psychophysiological study of compensatory tracking on a digital display. *Human Factors* 7: 457–472, 1965.

Bortolussi, M.R., Hart, S.G., and Shively, R.J. Measuring moment-to-moment pilot workload using synchronous presentations of secondary tasks in a motion-base trainer. Proceedings of the Fourth Symposium on Aviation Psychology. Columbus, OH: Ohio State University, 1987. Also published in *Aviation, Space, and Environmental Medicine* 60(2): 124–129, 1989.

Bortolussi, M.R., Kantowitz, B.H., and Hart, S.G. Measuring pilot workload in a motion base trainer: A comparison of four techniques. Proceedings of the Third Symposium on Aviation Psychology, 263–270, 1985.

Bortolussi, M.R., Kantowitz, B.H., and Hart, S.G. Measuring pilot workload in a motion base trainer. *Applied Ergonomics* 17: 278–283, 1986.

Brown, I.D., Tickner, A.H., and Simmonds, D.C.V. Interference between concurrent tasks of driving and telephoning. *Journal of Applied Psychology* 53: 419–424, 1969.

Damos, D. Residual attention as a predictor of pilot performance. *Human Factors* 20: 435–440, 1978.

Detweiler, M., and Lundy, D.H. Effects of single- and dual-task practice on acquiring dual-task skill. *Human Factors* 37(1): 193–211, 1995.

Drory, A. Effects of rest and secondary task on simulated truck-driving performance. *Human Factors* 27(2): 201–207, 1985.

Ellis, J.E. Analysis of temporal and attentional aspects of movement control. *Journal of Experimental Psychology* 99: 10–21, 1973.

Fisher, S. The microstructure of dual task interaction. 1. The patterning of main-task responses within secondary-task intervals. *Perception* 4: 267–290, 1975a.

Fisher, S. The microstructure of dual task interaction. 2. The effect of task instructions on attentional allocation and a model of attentional-switching. *Perception* 4: 459–474, 1975b.

Gawron, V.J. Performance effects of noise intensity, psychological set, and task type and complexity. *Human Factors* 24(2): 225–243, 1982.

Giroud, Y., Laurencelle, L., and Proteau, L. On the nature of the probe reaction-time task to uncover the attentional demands of movement. *Journal of Motor Behavior* 16: 442–459, 1984.

Glaser, Y.G., and Glaser, D.S. A comparison between a touchpad-controlled high forward Head-down Display (HF-HDD) and a touchscreen-controlled Head-down Display (HDD) for in-vehicle secondary tasks. Proceedings of the Human Factors and Ergonomics Society 59th Annual Meeting, 1387–1391, 2015.

Hicks, T.G., and Wierwille, W.W. Comparison of five mental workload assessment procedures in a moving-base during simulator. *Human Factors* 21: 129–143, 1979.

Israel, J.B., Chesney, G.L., Wickens, C.D., and Donchin, E. P300 and tracking difficulty: Evidence for multiple resources in dual-task performance. *Psychophysiology* 17: 259–273, 1980a.

Israel, J.B., Wickens, C.D., Chesney, G.L., and Donchin, E. The event-related brain potential as an index of display-monitoring workload. *Human Factors* 22: 211–224, 1980b.

Johnson, D.F., and Haygood, R.C. The use of secondary tasks in adaptive training. *Human Factors* 26(1): 105–108, 1984.

Kantowitz, B.H. Simulator evaluation of heavy-vehicle driver workload. Proceedings of the Human Factors and Ergonomics Society 39th Annual Meeting, 2: 1107–1111, 1995.

Kantowitz, B. H., Bortolussi, M.R., and Hart, S.G. Measuring pilot workload in a motion base simulator: III. Synchronous secondary task. *Proceedings of the Human Factors Society* 2: 834–837, 1987.

Kantowitz, B.H., Hart, S.G., and Bortolussi, M.R. Measuring pilot workload in a moving-base simulator: I. Asynchronous secondary choice-reaction time task. Proceedings of the 27th Annual Meeting of the Human Factors Society, 319–322, 1983.

Kantowitz, B.H., Hart, S.G., Bortolussi, M.R., Shively, R.J., and Kantowitz, S.C. Measuring pilot workload in a moving-base simulator: II. Building levels of workload. NASA 20th Annual Conference on Manual Control, vol. 2, 373–396, 1984.

Klapp, S.T., Kelly, P.A., Battiste, V., and Dunbar, S. Types of tracking errors induced by concurrent secondary manual task. Proceedings of the 20th Annual Conference on Manual Control, 299–304, 1984.

Klapp, S.T., Kelly, P.A., and Netick, A. Hesitations in continuous tracking induced by a concurrent discrete task. *Human Factors* 29(3): 327–337, 1987.

Krol, J.P. Variations in ATC-workload as a function of variations in cockpit workload. *Ergonomics* 14: 585–590, 1971.

Loeb, M., and Jones, P.D. Noise exposure, monitoring and tracking performance as a function of signal bias and task priority. *Ergonomics* 21(4): 265–272, 1978.

Logan, G.D. On the use of a concurrent memory load to measure attention and automaticity. *Journal of Experimental Psychology: Human Perception and Performance* 5: 189–207, 1970.

Lysaght, R.J., Hill, S.G., Dick, A.O., Plamondon, B.D., Linton, P.M., Wierwille, W.W., Zaklad, A.L., Bittner, A.C., and Wherry, R.J. Operator workload: Comprehensive review and evaluation of operator workload methodologies (Technical Report 851). Alexandria, VA: Army Research Institute for the Behavioral and Social Sciences, June 1989.

Schvaneveldt, R.W. Effects of complexity in simultaneous reaction time tasks. *Journal of Experimental Psychology* 81: 289–296, 1969.

Smith, M.C. Effect of varying channel capacity on stimulus detection and discrimination. *Journal of Experimental Psychology* 82: 520–526, 1969.

Wempe, T.E., and Baty, D.L. Human information processing rates during certain multiaxis tracking tasks with a concurrent auditory task. *IEEE Transactions on Man-Machine Systems* 9: 129–138, 1968.

2.2.3 Classification Secondary Task

General description – "The subject must judge whether symbol pairs are identical in form. For example, to match letters either on a physical level (AA) or on a name level (Aa)" (Lysaght et al., 1989, p. 233), or property (pepper is hot), or superset relation (an apple is a fruit). Cognitive processing requirements are discussed in Miller (1975).

Strengths and limitations – "Depending upon the requirements of the matching task, the task can impose demands on perceptual processes (physical match) and/or cognitive processes (name match or category match)" (Lysaght et al., 1989, p. 233). There have been differences in results based on the independent and dependent variables used. There have also been effects on the primary task.

Independent Variable. Beer et al. (1996) reported that performance on an aircraft classification task in the singular task mode did not predict performance in dual-task mode. In a similar experiment, Shaw et al. (2010) compared "painting" civilian vehicles while flying three Unmanned Aerial Vehicles (UAVs) in three control modes and two levels of task load. There was a significant effect of task load with more efficient painting in the low rather than the high task load condition. There was no significant effect of control mode.

Primary Task. Mastroianni and Schopper (1986) reported that performance on a secondary auditory classification task (easiest version – tone low or high, middle version – previous tone low or high, hardest version – tone prior to the previous tone low or high) was degraded as the difficulty of the task increased or the amount of force needed on the primary pursuit tracking task increased. Performance on the primary task degraded when the secondary task was present versus absent. However, Cao and Liu (2013) asked participants to judge if two sentences read to them had the same meaning or not while the participants were performing a simulated driving task. This task did not affect driving performance if driving was the primary task.

Dependent Variable. Damos (1985) reported that percentage correct scores for both single and dual-task performance were affected only by trial and not by either behavior pattern or pacing condition. Correct RT scores, however, were significantly related to trial and pacing by behavior pattern in the single-task condition and trial, behavior pattern, trial by pacing, and trial by behavior pattern in the dual-task condition.

Carter et al. (1986) reported that RT increased as the number of memory steps to verify a sentence increased. Slope was not a reliable measure of performance.

Data requirements – The following data are used to assess performance of this task: mean RT for physical match, mean RT for category match, number of errors for physical match, and number of errors for category match (Lysaght et al., 1989, p. 236).

Thresholds – Kobus et al. (1986) reported the following times to correctly classify one of five targets: visual = 224.6 seconds, auditory = 189.6 seconds, and multimodal (i.e., both visual and auditory) = 212.7 seconds. These conditions were not significantly different.

Sources

Beer, M.A., Gallaway, R.A., and Previc, R.H. Do individuals' visual recognition thresholds predict performance on concurrent attitude control flight tasks? *The International Journal of Aviation Psychology* 6(3): 273–297, 1996.

Cao, S., and Liu, Y. Gender factor in lane keeping and speech comprehension dual tasks. Proceedings of the Human Factors and Ergonomics Society 57th Annual Meeting, 1909–1913, 2013.

Carter, R.C., Krause, M., and Harbeson, M.M. Beware the reliability of slope scores for individuals. *Human Factors* 28(6): 673–683, 1986.

Damos, D. The relation between the type A behavior pattern, pacing, and subjective workload under single- and dual-task conditions. *Human Factors* 27(6): 675–680, 1985.

Kobus, D.A., Russotti, J., Schlichting, C., Haskell, G., Carpenter, S., and Wojtowicz, J. Multimodal detection and recognition performance of sonar operations. *Human Factors* 28(1): 23–29, 1986.

Lysaght, R.J., Hill, S.G., Dick, A.O., Plamondon, B.D., Linton, P.M., Wierwille, W.W., Zaklad, A.L., Bittner, A.C., and Wherry, R.J. Operator workload: Comprehensive review and evaluation of operator workload methodologies (Technical Report 851). Alexandria, VA: Army Research Institute for the Behavioral and Social Sciences, June 1989.

Mastroianni, G.R., and Schopper, A.W. Degradation of force-loaded pursuit tracking performance in a dual-task paradigm. *Ergonomics* 29(5): 639–647, 1986.

Miller, K. Processing capacity requirements for stimulus encoding. *Acta Psychologica* 39: 393–410, 1975.

Shaw, T., Emfield, A., Garcia, A., de Visser, E., Miller, C., Parasuraman, R., and Fern, L. Evaluating the benefits and potential costs of automation delegation for supervisory control of multiple UAVs. Proceedings of the Human Factors and Ergonomics Society 54th Annual Meeting, 1498–1502, 2010.

2.2.4 Cross-Adaptive Loading Secondary Task

General description – Cross-adaptive loading tasks are secondary tasks that the participant must perform only while primary task performance meets or exceeds a previously established performance criterion (Kelly and Wargo, 1967).

Strengths and limitations – Cross-adaptive loading tasks are less likely to degrade performance on the primary task but are intrusive and, as such, difficult to use in nonlaboratory settings.

Data requirements – A well-defined, quantifiable criterion for primary task performance as well as a method of monitoring this performance and cueing the participant when to perform the cross-adaptive loading task are all required.

Thresholds – Dependent on type of primary and cross-adaptive loading tasks being used.

Source

Kelly, C.R., and Wargo, M.J. Cross-adaptive operator loading tasks. *Human Factors* 9: 395–404, 1967.

2.2.5 Detection Secondary Task

General description – "The subject must detect a specific stimulus or event which may or may not be presented with alternative events. For example, to detect which of 4 lights is flickering. The subject is usually alerted by a warning signal (e.g., tone) before the occurrence of such events, therefore attention is required intermittently" (Lysaght et al., 1989, p. 233).

Strengths and limitations – "Such tasks are thought to impose demands on perceptual processes" (Lysaght et al., 1989, p. 233). Based on a review of

TABLE 2.2

References Listed by the Effect on Performance of Primary Tasks Paired with a Secondary Detection Task

	Primary Task			Secondary Task		
Type	Stable	Degraded	Enhanced	Stable	Degraded	Enhanced
Detection		Wickens et al. (1981)			Wickens et al. (1981)	
Driving		Moskovitch et al. (2010)[a]			Verwey (2000)[a]	
Memory					Shulman and Greenberg (1971)	
Tracking		Wickens et al. (1981)			Wickens et al. (1981)	

Source: From Lysaght et al. (1989, p. 246).
[a] Not included in Lysaght et al. (1989).

five studies in which detection was a secondary task, Lysaght et al. (1989) reported that, for dual-task pairings: performance of a primary classification task remained stable; and performance of tracking, memory, monitoring, and detection of primary tasks degraded. In all cases, performance of the secondary detection task degraded (see Table 2.2).

Data requirements – The following data are calculated for this task: mean RT for correct detections and number of correct detections.

Thresholds – Not stated.

Sources

Lysaght, R.J., Hill, S.G., Dick, A.O., Plamondon, B.D., Linton, P.M., Wierwille, W.W., Zaklad, A.L., Bittner, A.C., and Wherry, R.J. Operator workload: Comprehensive review and evaluation of operator workload methodologies (Technical Report 851). Alexandria, VA: Army Research Institute for the Behavioral and Social Sciences, June 1989.

Moskovitch, Y., Jeon, M., and Walker, B.N. Enhanced auditory menu cues on a mobile phone improve time-shared performance of a driving-like dual task. Proceedings of the Human Factors and Ergonomics Society 54th Annual Meeting, 1321–1325, 2010.

Shulman, H.G., and Greenberg, S.N. Perceptual deficit due to division of attention between memory and perception. *Journal of Experimental Psychology* 88: 171–176, 1971.

Verwey, W.B. On-line driver workload estimation. Effects of road situation and age on secondary task measures. *Ergonomics* 43(2): 187–209, 2000.

Wickens, C.D., Mountford, S.J., and Schreiner, W. Multiple resources, task-hemispheric integrity, and individual differences in time sharing. *Human Factors* 23: 211–229, 1981.

2.2.6 Distraction Secondary Task

General description – "The subject performs a task which is executed in a fairly automatic way such as counting aloud" (Lysaght et al., 1989, p. 233).

Strengths and limitations – "Such a task is intended to distract the subject in order to prevent the rehearsal of information that may be needed for the primary task" (Lysaght et al., 1989, p. 233).

Based on one study, Lysaght et al. (1989) reported degraded performance on a memory primary task when paired with a distraction secondary task. Drory (1985) reported significantly shorter brake RTs and fewer steering wheel reversals when a secondary distraction task (i.e., state the last two digits of the current odometer reading) was paired with a basic driving task in a simulator. There were no effects of the secondary distraction task on tracking error, number of brake responses, or control light responses.

Zeitlin (1995) used two auditory secondary tasks (delayed digit recall and random digit generation) while driving on the road. Performance on both tasks degraded as traffic density and average speed increased.

Data requirements – Not stated.
Thresholds – Not stated.

Sources

Drory, A. Effects of rest and secondary task on simulated truck-driving task performance. *Human Factors* 27(2): 201–207, 1985.

Lysaght, R.J., Hill, S.G., Dick, A.O., Plamondon, B.D., Linton, P.M., Wierwille, W.W., Zaklad, A.L., Bittner, A.C., and Wherry, R.J. Operator workload: Comprehensive review and evaluation of operator workload methodologies (Technical Report 851). Alexandria, VA: Army Research Institute for the Behavioral and Social Sciences, June 1989.

Zeitlin, L.R. Estimates of driver mental workload: A long-term field trial of two subsidiary tasks. *Human Factors* 37(3): 611–621, 1995.

2.2.7 Driving Secondary Task

General description – "The subject operates a driving simulator or actual motor vehicle" (Lysaght et al., 1989, p. 232).

Strengths and limitations – This "task involves complex psychomotor skills" (Lysaght et al., 1989, p. 232).

Brouwer et al. (1991) reported that older adults (mean age 64.4) were significantly worse than younger adults (mean age 26.1) in dual-task performance of compensatory lane tracking with a timed, self-paced visual analysis task.

Korteling (1994) did not find a significant difference in steering performance between single task (steering) versus dual task (addition of car following task) between young (21 to 34) and old (65 to 74 year old) drivers. There was, however, a 24% performance deterioration in car-following performance with the addition of a steering task.

Data requirements – The experimenter should be able to record: total time to complete a trial, number of acceleration rate changes, number of gear changes, number of footbrake operations, number of steering reversals, number of obstacles hit, high pass steering deviation, yaw deviation, and lateral deviation (Lysaght et al., 1989, p. 235).

Thresholds – Not stated.

Sources

Brouwer, W.H., Waterink, W., van Wolffelaar, P.C., and Rothengatten, T. Divided attention in experienced young and older drivers: Lane tracking and visual analysis in a dynamic driving simulator. *Human Factors* 33(5): 573–582, 1991.

Korteling, J.E. Effects of aging, skill modification, and demand alternation on multiple-task performance. *Human Factors* 36(1): 27–43, 1994.

Lysaght, R.J., Hill, S.G., Dick, A.O., Plamondon, B.D., Linton, P.M., Wierwille, W.W., Zaklad, A.L., Bittner, A.C., and Wherry, R.J. Operator workload: Comprehensive review and evaluation of operator workload methodologies (Technical Report 851). Alexandria, VA: Army Research Institute for the Behavioral and Social Sciences, June 1989.

2.2.8 Identification/Shadowing Secondary Task

General description – "The subject identifies changing symbols (digits and/or letters) that appear on a visual display by writing or verbalizing, or repeating a spoken passage as it occurs" (Lysaght et al., 1989, p. 233).

Strengths and limitations – "Such tasks are thought to impose demands on perceptual processes (i.e., attention)" (Lysaght et al., 1989, p. 233).

Based on nine studies with an identification secondary task, Lysaght et al. (1989) reported that performance: remained stable for an identification primary task; and degraded for tracking, memory, detection, driving, and spatial transformation primary tasks. Performance of the identification secondary task: remained stable for tracking and identification primary tasks; and degraded for monitoring, detection, driving, and spatial transformation primary tasks (see Table 2.3).

Wierwille and Connor (1983), however, reported that a digit shadowing task was not sensitive to variations in workload. Their task was control on a moving-base aircraft simulator. Workload was varied by manipulating pitch-stability level, wind-gust disturbance level, and crosswind velocity and direction.

Savage et al. (1978) evaluated the sensitivity of four dependent measures to workload manipulations (1, 2, 3, or 4 meters) to a primary monitoring task. The number of random digits spoken on the secondary task was the most sensitive to workload. The longest consecutive string of spoken digits and the number of triplets spoken were also significantly affected by workload. The longest interval between spoken responses, however, was not sensitive to workload manipulations of the primary task.

Data requirements – The following data are used for this task: number of words correct/minute, number of digits spoken, mean time interval between spoken digits, and number of errors of omission (Lysaght et al., 1989, p. 236).

Thresholds – Not stated.

TABLE 2.3
References Listed by the Effect on Performance of Primary Tasks Paired with a Secondary Identification Task

Type	Primary Task			Secondary Task		
	Stable	Degraded	Enhanced	Stable	Degraded	Enhanced
Detection		Price (1975)			Price (1975)	
Driving		Hicks and Wierwille (1979) Louie and Mouloua (2015)		Louie and Mouloua (2015)	Wierwille et al. (1977)	
Identification	Allport et al. (1972)			Allport et al. (1972)		
Memory		Mitsuda (1968)				
Monitoring					Savage et al. (1978)	
Spatial Transformation		Fournier and Stager (1976)			Fournier and Stager (1976)	
Tracking		Gabay and Merhav (1977)		Gabay and Merhav (1977)		

Source: From Lysaght et al. (1989, p. 246).

Sources

Allport, D.A., Antonis, B., and Reynolds, P. On the division of attention: A disproof of the single channel hypothesis. *Quarterly Journal of Experimental Psychology* 24: 225–235, 1972.
Fournier, B.A., and Stager, P. Concurrent validation of a dual-task selection test. *Journal of Applied Psychology* 5: 589–595, 1976.
Gabay, E., and Merhav, S.J. Identification of a parametric model of the human operator in closed-loop control tasks. *IEEE Transactions on Systems, Man, and Cybernetics.* SMC-7: 284–292, 1977.
Hicks, T.G., and Wierwille, W.W. Comparison of five mental workload assessment procedures in a moving-base driving simulator. *Human Factors* 21: 129–142, 1979.
Louie, J.F., and Mouloua, M. Individual differences in cognition as predictors of driving performance. Proceedings of the Human Factors and Ergonomics Society 59th Annual Meeting, 1540–1544, 2015.
Lysaght, R.J., Hill, S.G., Dick, A.O., Plamondon, B.D., Linton, P.M., Wierwille, W.W., Zaklad, A.L., Bittner, A.C., and Wherry, R.J. Operator workload: Comprehensive review and evaluation of operator workload methodologies (Technical Report 851). Alexandria, VA: Army Research Institute for the Behavioral and Social Sciences, June 1989.
Mitsuda, M. Effects of a subsidiary task on backward recall. *Journal of Verbal Learning and Verbal Behavior* 7: 722–725, 1968.
Price, D.L. The effects of certain gimbal orders on target acquisition and workload. *Human Factors* 20: 649–654, 1975.
Savage, R.E., Wierwille, W.W., and Cordes, R.E. Evaluating the sensitivity of various measures of operator workload using random digits as a secondary task. *Human Factors* 20: 649–654, 1978.
Wierwille, W.W., and Connor, S.A. Evaluation of 20 workload measures using a psychomotor task in a moving-base aircraft simulator. *Human Factors* 25(1): 1–16, 1983.
Wierwille, W.W., Gutmann, J.C., Hicks, T.G., and Muto, W.H. Secondary task measurement of workload as a function of simulated vehicle dynamics and driving conditions. *Human Factors* 19: 557–565, 1977.

2.2.9 Lexical Decision Secondary Task

General description – "Typically, the subject is briefly presented with a sequence of letters and must judge whether this letter sequence forms a word or a nonword" (Lysaght et al., 1989, p. 233).

Strengths and limitations – "This task is thought to impose heavy demands on semantic memory processes" (Lysaght et al., 1989, p. 233).

Data requirements – Mean RT for correct responses is used as data for this task.

Thresholds – Not stated.

Source

Lysaght, R.J., Hill, S.G., Dick, A.O., Plamondon, B.D., Linton, P.M., Wierwille, W.W., Zaklad, A.L., Bittner, A.C., and Wherry, R.J. Operator workload: Comprehensive review and evaluation of operator workload methodologies (Technical Report 851). Alexandria, VA: Army Research Institute for the Behavioral and Social Sciences, June 1989.

2.2.10 Memory Recall Secondary Task

General description – Participants are presented a series of numbers, letters, or words and must recall that series while doing a primary task. The series can be presented auditorily or visually.

Strengths and limitations – The task is easy to describe to participants and easy to score for number correct. He et al. (2013) presented numbers auditorily while participants were driving an automobile simulator. They reported that performance of the secondary task was associated with decreased lane variability but higher steering wheel reversal rate.

Data requirements – Collect the number correct and the number incorrect.

Thresholds – Zero to the maximum number of stimuli presented.

Source

He, J., McCarley, J.S., and Kramer, A.F. Lane keeping under cognitive load: Performance changes and mechanisms. *Human Factors* 56(2): 414–426, 2013.

2.2.11 Memory-Scanning Secondary Task

General description – These secondary tasks require a participant to memorize a list of letters, numbers, and/or shapes and then indicate whether a probe stimulus is a member of that set. Typically, there is a linear relation between the number of items in the memorized list and RT to the probe stimulus. A variation of this task involves recalling one of the letters based on its position in the presented set.

Strengths and limitations – The utility of memory tasks is dependent on how they are used: to predict performance, as single or dual task, and with what type of primary task they are paired.

Predict performance. Park and Lee (1992) reported memory tasks significantly predicted flight performance of pilot trainees.

Single versus dual task. The slope of the linear function may reflect memory-scanning rate. However, Carter et al. (1986) warned that the slope may

TABLE 2.4
References Listed by the Effect on Performance of Primary Tasks Paired with a Secondary Memory Task

	Primary Task			Secondary Task		
Type	Stable	Degraded	Enhanced	Stable	Degraded	Enhanced
Choice RT		Broadbent and Gregory (1965) Keele and Boies (1973)			Broadbent and Gregory (1965)	
Classification		Wickens et al. (1981)			Wickens et al. (1981)	
Detection		Wickens et al. (1981)			Wickens et al. (1981)	
Distraction		Broadbent and Heron (1962)				
Driving	Brown (1962, 1965, 1966) Brown and Poulton (1961) Kantowitz (1995) Wetherell (1981)	Richard et al. (2002; task was to search visual driving scene) Donmez et al. (2011) decreased turn signal use during on-road driving		Brown (1965)	Brown (1962, 1965, 1966) Brown and Poulton (1961) Wetherell (1981)	
Identification		Klein (1976)				
Memory		Broadbent and Heron (1962) Chow and Murdock (1975)		Shulman and Greenberg (1971)	Allport et al. (1972)	
Mental Math	Mandler and Worden (1973)			Mandler and Worden (1973)		

(Continued)

TABLE 2.4 (CONTINUED)
References Listed by the Effect on Performance of Primary Tasks Paired with a Secondary Memory Task

	Primary Task			Secondary Task		
Type	Stable	Degraded	Enhanced	Stable	Degraded	Enhanced
Monitoring	Chechile et al. (1979) Moskowitz and McGlothlin (1974)	Chiles and Alluisi (1979)		Chechile et al. (1979) Chiles and Alluisi (1979) Mandler and Worden (1973) Moskowitz and McGlothlin (1974)		
Problem Solving	Daniel et al. (1969)	Stager and Zufelt (1972)				
Tracking	Finkelman and Glass (1970) Zeitlin and Finkelman (1975)	Heimstra (1970) Huddleston and Wilson (1971) Noble et al. (1967) Trumbo and Milone (1971) Wickens and Kessel (1980) Wickens et al. (1981)	Tsang and Wickens (1984)	Noble et al. (1967) Trumbo and Milone (1971)	Finkelman and Glass (1970) Heimstra (1970) Huddleston and Wilson (1971) Tsang and Wickens (1984) Wickens and Kessel (1980) Wickens et al. (1981)	

Source: From Lysaght et al. (1989, p. 246).

Human Workload

be less reliable than the RTs used to calculate the slope. Measuring RTs, Fisk and Hodge (1992) reported no significant differences in RT in a single task performance of a memory-scanning task after 32 days without practice.

Type of primary task. Wierwille and Connor (1983), however, reported that a memory-scanning task was not sensitive to workload. Their primary task was control a moving-base aircraft simulator. Workload was varied by manipulating pitch-stability level, wind-gust disturbance level, and crosswind direction and velocity.

Based on 25 studies using a memory secondary task, Lysaght et al. (1989) reported that performance: remained stable on tracking, mental math, monitoring, problem solving, and driving primary tasks; degraded on tracking, choice RT, memory, monitoring, problem solving, detection, identification, classification, and distraction primary tasks; and improved on a tracking primary task. Performance of the memory secondary task remained stable with a tracking primary task; and degraded when paired with tracking, choice RT, memory, mental math, monitoring, detection, identification, classification, and driving primary tasks (see Table 2.4).

Data requirements – RT to items from the memorized list must be at asymptote to ensure that no additional learning will take place during the experiment. The presentation of the probe stimulus is intrusive and, thus, may be difficult to use in nonlaboratory settings.

Thresholds – 40 msec for RT.

Sources

Allport, D.A., Antonis, B., and Reynolds, P. On the division of attention: A disproof of the single channel hypothesis. *Quarterly Journal of Experimental Psychology* 24: 225–235, 1972.

Broadbent, D.E., and Gregory, M. On the interaction of S-R compatibility with other variables affecting reaction time. *British Journal of Psychology* 56: 61–67, 1965.

Broadbent, D.E., and Heron, A. Effects of a subsidiary task on performance involving immediate memory by younger and older men. *British Journal of Psychology* 53: 189–198, 1962.

Brown, I.D. Measuring the "spare mental capacity" of car drivers by a subsidiary auditory task. *Ergonomics* 5: 247–250, 1962.

Brown, I.D. A comparison of two subsidiary tasks used to measure fatigue in car drivers. *Ergonomics* 8: 467–473, 1965.

Brown, I.D. Subjective and objective comparisons of successful and unsuccessful trainee drivers. *Ergonomics* 9: 49–56, 1966.

Brown, I.D., and Poulton, E.C. Measuring the spare "mental capacity" of car drivers by a subsidiary task. *Ergonomics* 4: 35–40, 1961.

Carter, R.C., Krause, M., and Harbeson, M.M. Beware the reliability of slope scores for individuals. *Human Factors* 28(6): 673–683, 1986.

Chechile, R.A., Butler, K., Gutowski, W., and Palmer, E.A. Division of attention as a function of the number of steps, visual shifts and memory load. Proceedings of the 15th Annual Conference on Manual Control, 71–81, 1979.

Chiles, W.D., and Alluisi, E.A. On the specification of operator or occupational workload performance-measurement methods. *Human Factors* 21: 515–528, 1979.

Chow, S.L., and Murdock, B.B. The effect of a subsidiary task on iconic memory. *Memory and Cognition* 3: 678–688, 1975.

Daniel, J., Florek, H., Kosinar, V., and Strizenec, M. Investigation of an operator's characteristics by means of factorial analysis. *Studia Psychologica* 11: 10–22, 1969.

Donmez, B., Reimer, B., Mehler, B., Lavalliere, M., and Coughlin, J.F. A pilot investigation of the impact of cognitive demand on turn signal use during lane changes in actual highway conditions across multiple age groups. Proceedings of the Human Factors and Ergonomics Society 55th Annual Meeting, 1874–1878, 2011.

Finkelman, J.M., and Glass, D.C. Reappraisal of the relationship between noise and human performance by means of a subsidiary task measure. *Journal of Applied Psychology* 54: 211–213, 1970.

Fisk, A.D., and Hodge, K.A. Retention of trained performance in consistent mapping search after extended delay. *Human Factors* 34(2): 147–164, 1992.

Heimstra, N.W. The effects of "stress fatigue" on performance in a simulated driving situation. *Ergonomics* 13: 209–218, 1970.

Huddleston, J.H.F., and Wilson, R.V. An evaluation of the usefulness of four secondary tasks in assessing the effect of a lag in simulated aircraft dynamics. *Ergonomics* 14: 371–380, 1971.

Kantowitz, B.H. Simulator evaluation of heavy-vehicle driver workload. Proceedings of the Human Factors and Ergonomics Society 39th Annual Meeting, 1107–1111, 1995.

Keele, S.W., and Boies, S.J. Processing demands of sequential information. *Memory and Cognition* 1: 85–90, 1973.

Klein, G.A. Effect of attentional demands on context utilization. *Journal of Educational Psychology* 68: 25–31, 1976.

Lysaght, R.J., Hill, S.G., Dick, A.O., Plamondon, B.D., Linton, P.M., Wierwille, W.W., Zaklad, A.L., Bittner, A.C., and Wherry, R.J. Operator workload: Comprehensive review and evaluation of operator workload methodologies (Technical Report 851). Alexandria, VA: Army Research Institute for the Behavioral and Social Sciences, June 1989.

Mandler, G., and Worden, P.E. Semantic processing without permanent storage. *Journal of Experimental Psychology* 100: 277–283, 1973.

Moskowitz, H., and McGlothlin, W. Effects of marijuana on auditory signal detection. *Psychopharmacologia* 40: 137–145, 1974.

Noble, M., Trumbo, D., and Fowler, F. Further evidence on secondary task interference in tracking. *Journal of Experimental Psychology* 73: 146–149, 1967.

Park, K.S., and Lee, S.W. A computer-aided aptitude test for predicting flight performance of trainees. *Human Factors* 34(2): 189–204, 1992.

Richard, C.M., Wright, R.D., Ee, C., Prime, S.L., Shimizu, Y., and Vavrik, J. Effect of a concurrent auditory task on visual search performance in a driving-related image-flicker task. *Human Factors* 44(1): 108–119, 2002.

Shulman, H.G., and Greenberg, S.N. Perceptual deficit due to division of attention between memory and perception. *Journal of Experimental Psychology* 88: 171–176, 1971.

Stager, P., and Zufelt, K. Dual-task method in determining load differences. *Journal of Experimental Psychology* 94: 113–115, 1972.

Trumbo, D., and Milone, F. Primary task performance as a function of encoding, retention, and recall in a secondary task. *Journal of Experimental Psychology* 91: 273–279, 1971.

Tsang, P.S., and Wickens, C.D. The effects of task structures on time-sharing efficiency and resource allocation optimality. Proceedings of the 20th Annual Conference on Manual Control, 305–317, 1984.

Wetherell, A. The efficacy of some auditory-vocal subsidiary tasks as measures of the mental load on male and female drivers. *Ergonomics* 24: 197–214, 1981.

Wickens, C.D., and Kessel, C. Processing resource demands of failure detection in dynamic systems. *Journal of Experimental Psychology: Human Perception and Performance* 6: 564–577, 1980.

Wickens, C.D., Mountford, S.J., and Schreiner, W. Multiple resources, task-hemispheric integrity, and individual differences in time sharing. *Human Factors* 23: 211–229, 1981.

Wierwille, W.W., and Connor, S.A. Evaluation of 20 workload measures using a psychomotor task in a moving-base aircraft simulator. *Human Factors* 25(1): 1–16, 1983.

Zeitlin, L.R., and Finkelman, J.M. Research note: Subsidiary task techniques of digit generation and digit recall indirect measures of operator loading. *Human Factors* 17: 218–220, 1975.

2.2.12 Mental Mathematics Secondary Task

General description – Participants are asked to perform arithmetic operations (i.e., addition, subtraction, multiplication, and division) on sets of visually or aurally presented digits.

Strengths and limitations – The major strength of this workload measure is its ability to discriminate between good and poor performers and high and low workload. For example, Ramacci and Rota (1975) required pilot applicants to perform progressive subtraction during their initial flight training. They reported that the number of subtractions performed increased while the percent of errors decreased with flight experience. Further, successful applicants performed more subtractions and had a lower percentage of errors than those applicants who were not accepted.

Green and Flux (1977) required pilots to add the digit three to aurally presented digits during a simulated flight. They reported increased performance time of the secondary task as the workload associated with the primary task increased. Huddleston and Wilson (1971) asked pilots to determine if digits were odd or even, their sum was odd or even, two consecutive digits were the same or different, or every other digit was the same or different. Again, secondary task performance discriminated between high and low workload on the primary task.

The major disadvantage of this secondary task is intrusion into the primary task. Andre et al. (1995) reported greater root mean square error (rmse)

TABLE 2.5
References Listed by the Effect on Performance of Primary Tasks Paired with a Secondary Mental Math Task

Type	Primary Task Stable	Primary Task Degraded	Primary Task Enhanced	Secondary Task Stable	Secondary Task Degraded	Secondary Task Enhanced
Choice RT		Chiles and Jennings (1970) Fisher (1975) Keele (1967)			Fisher (1975) Keele (1967) Schouten et al. (1962)	
Detection		Jaschinski (1982)			Jaschinski (1982)	
Driving	Brown and Poulton (1961) Wetherell (1981)			Verwey (2000)[a]	Brown and Poulton (1961) Wetherell (1981)	
Memory		Roediger et al. (1977) Silverstein and Glanzer (1971)				
Monitoring		Chiles and Jennings (1970) Kahneman et al. (1967)			Chiles and Jennings (1970) Kahneman et al. (1967)	
Simple RT		Chiles and Jennings (1970) Green and Flux (1977)[a] Wierwille and Connor (1983)[a]			Green and Flux (1977)[a]	
Simulated Flight Task	Green and Flux (1977)[a] Wierwille and Connor (1983)[a]	Andre et al. (1995)[a]			Green and Flux (1977)[a]	
Tapping	Kantowitz and Knight (1974)				Kantowitz and Knight (1974, 1976)	
Tracking	Huddleston and Wilson (1971)	Bahrick et al. (1954) Chiles and Jennings (1970) Heimstra (1970) McLeod (1973) Wickens et al. (1981)	Kramer et al. (1984)	Bahrick et al. (1954) Heimstra (1970)	Huddleston and Wilson (1971) McLeod (1973) Wickens et al. (1981)	Chiles and Jennings (1970)

Source: From Lysaght et al. (1989) p. 247.
[a] Not included in Lysaght et al. (1989).

in roll, pitch, and yaw in a primary simulated flight task when paired with a mental mathematics secondary task (i.e., fuel range). Harms (1986) reported similar results for a driving task.

Mental mathematics tasks have also been used in the laboratory. For example, Kramer et al. (1984) reported a significant increase in tracking error on the primary task when the secondary task was counting flashes. Damos (1985) required participants to calculate the absolute difference between the digit currently presented visually and the digit that had preceded it. In the single-task condition, percentage correct scores were significantly related to trial. Correct RT scores were related to trial and trial by pacing condition. In the dual-task condition, percentage correct scores were not significantly related to trial, behavior pattern, or pacing condition. Correct RT in the dual-task condition, however, was related to trial, trial by pacing, and trial by behavior pattern.

Based on 15 studies that used a mental math secondary task, Lysaght et al. (1989) reported that performance remained the same for tracking, driving, and tapping primary tasks and degraded for tracking, choice RT, memory, monitoring, simple RT, and detection primary tasks. Performance of the mental math secondary task remained stable with a tracking primary task; degraded with tracking, choice RT, monitoring, detection, driving, and tapping primary tasks; and improved with tracking primary task (see Table 2.5).

Data requirements – The following data are calculated: number of correct responses, mean RT for correct responses, and number of incorrect responses (Lysaght et al., 1989, p. 235). The researcher should compare primary task performance with and without a secondary task to ensure that participants are not sacrificing primary task performance to enhance secondary task performance.

Thresholds – Not stated.

Sources

Andre, A.D., Heers, S.T., and Cashion, P.A. Effects of workload preview on task scheduling during simulated instrument flight. *International Journal of Aviation Psychology* 5(1): 5–23, 1995.

Bahrick, H.P., Noble, M., and Fitts, P.M. Extra-task performance as a measure of learning task. *Journal of Experimental Psychology* 4: 299–302, 1954.

Brown, I.D., and Poulton, E.C. Measuring the spare "mental capacity" of car drivers by a subsidiary task. *Ergonomics* 4: 35–40, 1961.

Chiles, W.D., and Jennings, A.E. Effects of alcohol on complex performance. *Human Factors* 12: 605–612, 1970.

Damos, D. The relation between the Type A behavior pattern, pacing, and subjective workload under single- and dual-task conditions. *Human Factors* 27(6): 675–680, 1985.

Fisher, S. The microstructure of dual task interaction. 1. The patterning of main task response within secondary-task intervals. *Perception* 4: 267–290, 1975.

Green, R., and Flux, R. Auditory communication and workload. Proceedings of NATO Advisory Group for Aerospace Research and Development Conference on Methods to Assess Workload (AGARD-CPP-216), A4-1–A4-8, 1977.

Harms, L. Drivers' attentional response to environmental variations: A dual-task real traffic study. In A.G. Gale, M.H. Freeman, C.M. Haslegrave, P. Smith, and S.P. Taylor (Eds.) *Vision in Vehicles* (pp. 131–138). Amsterdam: North Holland, 1986.

Heimstra, N.W. The effects of "stress fatigue" on performance in a simulated driving situation. *Ergonomics* 13: 209–218, 1970.

Huddleston, J.H.F., and Wilson, R.V. An evaluation of the usefulness of four secondary tasks in assessing the effect of a log in simulated aircraft dynamics. *Ergonomics* 14: 371–380, 1971.

Jaschinski, W. Conditions of emergency lighting. *Ergonomics* 25: 363–372, 1982.

Kahneman, D., Beatty, J., and Pollack, I. Perceptual deficit during a mental task. *Science* 157: 218–219, 1967.

Kantowitz, B.H., and Knight, J.L. Testing tapping time-sharing. *Journal of Experimental Psychology* 103: 331–336, 1974.

Kantowitz, B.H., and Knight, J.L. Testing tapping time sharing: II. Auditory secondary task. *Acta Psychologica* 40: 343–362, 1976.

Keele, S.W. Compatibility and time-sharing in serial reaction time. *Journal of Experimental Psychology* 75: 529–539, 1967.

Kramer, A.F., Wickens, C.D., and Donchin, E. Performance and enhancements under dual-task conditions. Annual Conference on Manual Control, 21–35, 1984.

Lysaght, R.J., Hill, S.G., Dick, A.O., Plamondon, B.D., Linton, P.M., Wierwille, W.W., Zaklad, A.L., Bittner, A.C., and Wherry, R.J. Operator workload: Comprehensive review and evaluation of operator workload methodologies (Technical Report 851). Alexandria, VA: Army Research Institute for the Behavioral and Social Sciences, June 1989.

McLeod, P.D. Interference of "attend to and learn" tasks with tracking. *Journal of Experimental Psychology* 99: 330–333, 1973.

Ramacci, C.A., and Rota, P. Flight fitness and psycho-physiological behavior of applicant pilots in the first flight missions. Proceedings of NATO Advisory Group for Aerospace Research and Development (AGARD 153), B8, 1975.

Roediger, H.L., Knight, J.L., and Kantowitz, B.H. Inferring delay in short-term memory: The issue of capacity. *Memory and Cognition* 5: 167–176, 1977.

Schouten, J.F., Kalsbeek, J.W.H., and Leopold, F.F. On the evaluation of perceptual and mental load. *Ergonomics* 5: 251–260, 1962.

Silverstein, C., and Glanzer, M. Concurrent task in free recall: Differential effects of LTS and STS. *Psychonomic Science* 22: 367–368, 1971.

Verwey, W.B. On-line driver workload estimation. Effects of road situation and age on secondary task measures. *Ergonomics* 43(2): 187–209, 2000.

Wetherell, A. The efficacy of some auditory-vocal subsidiary tasks as measures of the mental load on male and female drivers. *Ergonomics* 24: 197–214, 1981.

Wickens, C.D., Mountford, S.J., and Schreiner, W. Multiple resources, task-hemispheric integrity, and individual differences in time-sharing. *Human Factors* 23: 211–229, 1981.

Wierwille, W.W., and Connor, S. Evaluation of 20 workload measures using a psychomotor task in a moving base aircraft simulator. *Human Factors* 25: 1–16, 1983.

2.2.13 Michon Interval Production Secondary Task

General description – "The Michon paradigm of interval production requires the participant to generate a series of regular time intervals by executing a motor response (i.e., a single finger tap [every] 2 sec.). No sensory input is required." (Lysaght et al., 1989, p. 233).

Strengths and limitations – "This task is thought to impose heavy demand on motor output/response resources. It has been demonstrated with high demand primary tasks that participants exhibit irregular or variable tapping rates" (Lysaght et al., 1989, p. 233). Crabtree et al. (1984) reported that scores on this secondary task discriminated the workload of three switch-setting tasks. Johannsen et al. (1976) reported similar results for autopilot evaluation in a fixed-based flight simulator. Wierwille et al. (1985b), however, reported that tapping regularity was not affected by variations in the difficulty of a mathematical problem to be solved during simulated flight.

In a surgical training application, Grant et al. (2013) provided guidelines for the use of interval production as a measure of workload: 1) interval between three and 30 seconds, 2) match response modality to experimental constraints, 3) calculate both percentage absolute error and coefficient of variation, 4) collect a minimum of five samples per trial, 5) provide at least with practice trials no primary task and one with a primary task, 6) do not encourage/discourage strategies, and 7) instruct participants to perform both primary and interval production task with the primary being more critical. Further, Drury (1972) recommended a correlation to provide a non-dimensional measure.

Based on a review of six studies in which the Michon Interval Production task was the secondary task, Lysaght et al. (1989) stated that, in dual-task pairings, performance of flight simulation and driving primary tasks remained stable; performance of monitoring, problem solving, detection, psychomotor, Sternberg, tracking, choice RT, memory, and mental math primary tasks degraded; and performance of simple RT primary task improved. In these same pairings, performance of the Michon Interval Production task remained stable with monitoring, Sternberg, flight simulation, and memory primary tasks; and degraded with problem solving, simple RT, detection, psychomotor, flight simulation, driving, tracking, choice RT, and mental math primary tasks (see Table 2.6).

Data requirements – Michon (1966) stated that a functional description of behavior is needed for the technique. The following data are calculated: mean interval per trial, standard deviation of interval per trial, and sum of differences between successive intervals per minute of total time (Lysaght et al., 1989, p. 235).

Thresholds – Not stated.

TABLE 2.6
References Listed by the Effect on Performance of Primary Tasks Paired with a Secondary Michon Interval Production Task

Type	Primary Task Stable	Primary Task Degraded	Primary Task Enhanced	Secondary Task Stable	Secondary Task Degraded	Secondary Task Enhanced
Choice RT		Michon (1964)			Michon (1964)	
Detection		Michon (1964)			Michon (1964)	
Driving	Brown (1967)	Brown et al. (1967)[a]			Brown (1967) Brown et al. (1967)[a]	
Flight Simulation	Wierwille et al. (1985a)			Wierwille et al. (1985b)	Wierwille et al. (1985b)	
Memory		Roediger et al. (1977)		Roediger et al. (1977)		
Mental Math		Michon (1964)			Michon (1964)	
Monitoring		Shingledecker et al. (1983)		Shingledecker et al. (1983)		
Problem Solving		Michon (1964)			Michon (1964)	
Psychomotor		Michon (1964)			Michon (1964)	
Simple RT			Vroon (1973)		Vroon (1973)	
Sternberg		Shingledecker et al. (1983)		Shingledecker et al. (1983)		
Tracking		Shingledecker et al. (1983)			Shingledecker et al. (1983)	

Source: From Lysaght et al. (1989, p. 245).
[a] Not included in Lysaght et al. (1989).

Sources

Brown, I.D. Measurement of control skills, vigilance, and performance on a subsidiary task during twelve hours of car driving. *Ergonomics* 10: 665–673, 1967.

Brown, I.D., Simmonds, D.C.V., and Tickner, A.H. Measurement of control skills, vigilance, and performance on a subsidiary task during 12 hours of car driving. *Ergonomics* 10: 655–673, 1967.

Crabtree, M.S., Bateman, R.P., and Acton, W. Benefits of using objective and subjective workload measures. Proceedings of the 28th Annual Meeting of the Human Factors Society, 950–953, 1984.

Drury, C.G. Note on Michon's measure of tapping irregularity. *Ergonomics* 15(2): 195–197, 1972.

Grant, R.C., Carswell, C.M., Lio, C.H., and Seales, W.B. Measuring surgeons' mental workload with a time-based secondary task. *Ergonomics In Design* 21(1): 4–11, 2013.

Johannsen, G., Pfendler, C., and Stein, W. Human performance and workload in simulated landing approaches with autopilot failures. In N. Moray (Ed.) *Mental Workload, Its Theory and Measurement* (pp. 101–104). New York: Plenum Press, 1976.

Lysaght, R.J., Hill, S.G., Dick, A.O., Plamondon, B.D., Linton, P.M., Wierwille, W.W., Zaklad, A.L., Bittner, A.C., and Wherry, R.J. Operator workload: Comprehensive review and evaluation of operator workload methodologies (Technical Report 851). Alexandria, VA: Army Research Institute for the Behavioral and Social Sciences, June 1989.

Michon, J.A. A note on the measurement of perceptual motor load. *Ergonomics* 7: 461–463, 1964.

Michon, J.A. Tapping regularity as a measure of perceptual motor load. *Ergonomics* 9(5): 401–412, 1966.

Roediger, H.L., Knight, J.L., and Kantowitz, B.H. Inferring decay in short-term memory: The issue of capacity. *Memory and Cognition* 5: 167–176, 1977.

Shingledecker, C.A., Acton, W., and Crabtree, M.S. Development and application of a criterion task set for workload metric evaluation (SAE Technical Paper No. 831419). Warrendale, PA: Society of Automotive Engineers, 1983.

Vroon, P.A. Tapping rate as a measure of expectancy in terms of response and attention limitation. *Journal of Experimental Psychology* 101: 183–185, 1973.

Wierwille, W.W., Casali, J.G., Connor, S.A., and Rahimi, M. Evaluation of the sensitivity and intrusion of mental workload estimation techniques. In W. Roner (Ed.) *Advances in Man-Machine Systems Research*, vol. 2 (pp. 51–127). Greenwich, CT: J.A.I. Press, 1985a.

Wierwille, W.W., Rahimi, M., and Casali, J.G. Evaluation of 16 measures of mental workload using a simulated flight task emphasizing mediational activity. *Human Factors* 27(5): 489–502, 1985b.

2.2.14 Monitoring Secondary Task

General description – Participants are asked to respond either manually or verbally to the onset of visual or auditory stimuli. Both the time to respond and the accuracy of the response have been used as workload measures.

Strengths and limitations – The major advantage of the monitoring-task technique is its relevance to system safety. It is also able to discriminate among levels of automation and workload. For example, Anderson and Toivanen (1970) used a force-paced digit-naming task as a secondary task to investigate the effects of varying levels of automation in a helicopter simulator. Bortolussi et al. (1986) reported significant differences in two- and four-choice visual RT in easy and difficult flight scenarios. Bortolussi et al. (1987) reported significantly longer RTs in a four-choice RT task during a high-difficulty scenario than during a low-difficulty one. Lokhande and Reynolds (2012) used monitoring progress of flights and updating flight strips in an air traffic tower as a secondary measure of workload. Twelve controllers "shadowed" the live operations at the Dallas Fort Worth Tower. The primary task was issuing verbal commands to pilots. The dependent variable was communication gap time. Outside of aviation, Brown (1969) studied it in relation to flicker.

Based on the results of 36 studies that included a secondary monitoring task, Lysaght et al. (1989) reported that performance remained stable on tracking, choice RT, memory, mental math, problem solving, identification, and driving primary tasks; degraded on tracking, choice RT, memory, monitoring, detection, and driving primary tasks; and improved on a monitoring primary task. Performance of the monitoring secondary task remained stable when paired with tracking, memory, monitoring, flight simulation, and driving primary tasks; degraded when paired with tracking, choice RT, mental math, monitoring, problem solving, detection, identification and driving primary tasks; and improved when paired with tracking and driving primary tasks (see Table 2.7).

Data requirements – The experimenter should calculate: number of correct detections, number of incorrect detections, number of errors of omission, mean RT for correct detections, and mean RT for incorrect detections (Lysaght et al., 1989, p. 235). The Knowles (1963) guidelines are appropriate in selecting a vigilance task. In addition, the modality of the task must not interfere with performance of the primary task, for example, requiring a verbal response while a pilot is communicating with Air Traffic Control or other crewmembers.

Thresholds – Not stated.

Sources

Anderson, P.A., and Toivanen, M.L. Effects of varying levels of autopilot assistance and workload on pilot performance in the helicopter formation flight mode (Technical Report JANAIR 680610). Washington, D.C.: Office of Naval Research, March 1970.

TABLE 2.7
References Listed by the Effect on Performance of Primary Tasks Paired with a Secondary Monitoring Task

Type	Primary Task				Secondary Task		
	Stable	Degraded	Enhanced	Stable	Degraded	Enhanced
Choice RT	Boggs and Simon (1968)	Hilgendorf (1967)			Hilgendorf (1967)	
Detection		Dewar et al. (1976) Tyler and Halcomb (1974)			Tyler and Halcomb (1974)	
Driving	Brown (1962, 1967) Hoffman and Jorbert (1966) Wetherell (1981)	Brown (1965)		Hoffman and Jorbert (1966)	Brown (1962, 1965)	Brown (1967)
Flight simulation				Soliday and Schohan (1965)		
Identification	Dornic (1980)				Dornic (1980) Chiles et al. (1979)	
Memory	Tyler and Halcomb (1974)	Chow and Murdock (1975) Lindsay and Norman (1969) Mitsuda (1968)		Lindsay and Norman (1969)		
Mental Math	Dornic (1980)				Chiles et al. (1979) Dornic (1980)	
Monitoring		Chechile et al. (1979) Fleishman (1965) Goldstein and Dorfman (1978) Hohmuth (1970) Long (1976) Stager and Muter (1971)	McGrath (1965)	Stager and Muter (1971)	Chechile et al. (1979) Hohmuth (1970) Long (1976)	

(*Continued*)

TABLE 2.7 (CONTINUED)
References Listed by the Effect on Performance of Primary Tasks Paired with a Secondary Monitoring Task

	Primary Task			Secondary Task		
Type	Stable	Degraded	Enhanced	Stable	Degraded	Enhanced
Problem Solving	Wright et al. (1974)				Chiles et al. (1979) Wright et al. (1974)	
Tracking	Bell (1978) Figarola and Billings (1966) Gabriel and Burrows (1968) Huddleston and Wilson (1971) Kelley and Wargo (1967) Kyriakides and Leventhal (1977) Schori and Jones (1975)	Bergeron (1968) Heimstra (1970) Herman (1965) Kramer et al. (1984) Malmstrom et al. (1983) Monty and Ruby (1965) Putz and Rothe (1974)		Figarola and Billings (1966) Kramer et al. (1984) Malmstrom et al. (1983)	Bell (1978) Bergeron (1968) Gabriel and Burrows (1968) Herman (1965) Huddleston and Wilson (1971) Kelley and Wargo (1967) Kyriakides and Leventhal (1977) Monty and Ruby (1965) Putz and Rothe (1974) Schori and Jones (1975)	Heimstra (1970)

Source: From Lysaght et al. (1989, p. 246).
[a] Not included in Lysaght et al. (1989).

Bell, P.A. Effects of noise and heat stress on primary and subsidiary task performance. *Human Factors* 20: 749–752, 1978.

Bergeron, H.P. Pilot response in combined control tasks. *Human Factors* 10: 277–282, 1968.

Boggs, D.H., and Simon, J.R. Differential effect of noise on tasks of varying complexity. *Journal of Applied Psychology* 52: 148–153, 1968.

Bortolussi, M.R., Hart, S.G., and Shively, R.J. Measuring moment-to-moment pilot workload using synchronous presentations of secondary tasks in a motion-base trainer. Proceedings of the 4th Symposium on Aviation Psychology, 651–657, 1987.

Bortolussi, M.R., Kantowitz, B.H., and Hart, S.G. Measuring pilot workload in a motion base trainer: A comparison of four techniques. *Applied Ergonomics* 17: 278–283, 1986.

Brown, I.D. Measuring the "spare mental capacity" of car drivers by a subsidiary auditorssy task. *Ergonomics* 5: 247–250, 1962.

Brown, I.D. A comparison of two subsidiary tasks used to measure fatigue in car drivers. *Ergonomics* 8: 467–473, 1965.

Brown, I.D. Measurement of control skills, vigilance, and performance on a subsidiary task during twelve hours of car driving. *Ergonomics* 10: 665–673, 1967.

Brown, J.L. Flicker and intermittent stimulation. In C.H. Graham (Ed.) *Vision and Visual Perception*. New York: Wiley, 1969.

Chechile, R.A., Butler, K., Gutowski, W., and Palmer, E.A. Division of attention as a function of the number of steps, visual shifts, and memory load. Proceedings of the 15th Annual Conference on Manual Control, 71–81, 1979.

Chiles, W.D., Jennings, A.E., and Alluisi, E.C. Measurement and scaling of workload in complex performance. *Aviation, Space, and Environmental Medicine* 50: 376–381, 1979.

Chow, S.L, and Murdock, B.B. The effect of a subsidiary task on iconic memory. *Memory and Cognition* 3: 678–688, 1975.

Dewar, R.E., Ellis, J.E., and Mundy, G. Reaction time as an index of traffic sign perception. *Human Factors* 18: 381–392, 1976.

Dornic, S. Language dominance, spare capacity and perceived effort in bilinguals. *Ergonomics* 23: 369–377, 1980.

Figarola, T.R., and Billings, C.E. Effects of meprobamate and hypoxia on psychomotor performance. *Aerospace Medicine* 37: 951–954, 1966.

Fleishman, E.A. The prediction of total task performance from prior practice on task components. *Human Factors* 7: 18–27, 1965.

Gabriel, R.F., and Burrows, A.A. Improving time-sharing performance of pilots through training. *Human Factors* 10: 33–40, 1968.

Goldstein, I.L., and Dorfman, P.W. Speed and load stress as determinants of performance in a time sharing task. *Human Factors* 20: 603–609, 1978.

Heimstra, N.W. The effects of "stress fatigue" on performance in a simulated driving situation. *Ergonomics* 13: 209–218, 1970.

Herman, L.M. Study of the single channel hypothesis and input regulation within a continuous, simultaneous task situation. *Quarterly Journal of Experimental Psychology* 17: 37–46, 1965.

Hilgendorf, E.L. Information processing practice and spare capacity. *Australian Journal of Psychology* 19: 241–251, 1967.

Hoffman, E.R., and Jorbert, P.N. The effect of changes in some vehicle handling variables on driver steering performance. *Human Factors* 8: 245–263, 1966.

Hohmuth, A.V. Vigilance performance in a bimodal task. *Journal of Applied Psychology* 54: 520–525, 1970.

Huddleston, J.H.F., and Wilson, R.V. An evaluation of the usefulness of four secondary tasks in assessing the effect of a lag in simulated aircraft dynamics. *Ergonomics* 14: 371–380, 1971.

Kelley, C.R., and Wargo, M.J. Cross-adaptive operator loading tasks. *Human Factors* 9: 395–404, 1967.

Knowles, W.B. Operator loading tasks. *Human Factors* 5: 151–161, 1963.

Kramer, A.F., Wickens, C.D., and Donchin, E. Performance enhancements under dual-task conditions. Proceedings of the 20th Annual Conference on Manual Control, 21–35, 1984.

Kyriakides, K., and Leventhal, H.G. Some effects of intrasound on task performance. *Journal of Sound and Vibration* 50: 369–388, 1977.

Lindsay, P.H., and Norman, D.A. Short-term retention during a simultaneous detection task. *Perception and Psychophysics* 5: 201–205, 1969.

Lokhande, K., and Reynolds, H.J.D. Cognitive workload and visual attention analyses of the air traffic control Tower Flight Data Manager (TFDM) prototype demonstration. Proceedings of the Human Factors and Ergonomics 56th Annual Meeting, 105–109, 2012.

Long, J. Effect on task difficulty on the division of attention between nonverbal signals: Independence or interaction? *Quarterly Journal of Experimental Psychology* 28: 179–193, 1976.

Lysaght, R.J., Hill, S.G., Dick, A.O., Plamondon, B.D., Linton, P.M., Wierwille, W.W., Zaklad, A.L., Bittner, A.C., and Wherry, R.J. Operator workload: Comprehensive review and evaluation of operator workload methodologies (Technical Report 851). Alexandria, VA: Army Research Institute for the Behavioral and Social Sciences, June 1989.

Malmstrom, F.V., Reed, L.E., and Randle, R.J. Restriction of pursuit eye movement range during a concurrent auditory task. *Journal of Applied Psychology* 68: 565–571, 1983.

McGrath, J.J. Performance sharing in an audio-visual vigilance task. *Human Factors* 7: 141–153, 1965.

Mitsuda, M. Effects of a subsidiary task on backward recall. *Journal of Verbal Learning and Verbal Behavior* 7: 722–725, 1968.

Monty, R.A., and Ruby, W.J. Effects of added workload on compensatory tracking for maximum terrain following. *Human Factors* 7: 207–214, 1965.

Putz, V.R., and Rothe, R. Peripheral signal detection and concurrent compensatory tracking. *Journal of Motor Behavior* 6: 155–163, 1974.

Schori, T.R., and Jones, B.W. Smoking and workload. *Journal of Motor Behavior* 7: 113–120, 1975.

Soliday, S.M., and Schohan, B. Task loading of pilots in simulated low-altitude high-speed flight. *Human Factors* 7: 45–53, 1965.

Stager, P., and Muter, P. Instructions and information processing in a complex task. *Journal of Experimental Psychology* 87: 291–294, 1971.

Tyler, D.M., and Halcomb, C.G. Monitoring performance with a time-shared encoding task. *Perceptual and Motor Skills* 38: 383–386, 1974.

Wetherell, A. The efficacy of some auditory vocal subsidiary tasks as measures of the mental load on male and female drivers. *Ergonomics* 24: 197–214, 1981.

Wright, P., Holloway, C.M., and Aldrich, A.R. Attending to visual or auditory verbal information while performing other concurrent tasks. *Quarterly Journal of Experimental Psychology* 26: 454–463, 1974.

2.2.15 Multiple Task Performance Battery of Secondary Tasks

General description – The Multiple Task Performance Battery (MTPB) requires participants to time-share three or more of the following tasks: (1) light and dial monitoring, (2) mental math, (3) pattern discrimination, (4) target identification, (5) group problem solving, and (6) two-dimensional compensatory tracking. The monitoring tasks are used as secondary tasks and performance associated with these tasks as measures of workload.

Strengths and limitations – Increasing the number of tasks being time-shared does increase the detection time associated with the monitoring task. The MTPB may be difficult to implement in nonlaboratory settings. Lysaght et al. (1989) reported the results of one study (Alluisi and Morgan, 1971) in which the MTPB was paired with itself. In dual-task performance, performance of both the primary and the secondary MTPB tasks degraded.

Data requirements – The MTPB requires individual programming and analysis of six tasks as well as coordination among them during the experiment.

Thresholds – Not stated.

Sources

Alluisi, E.A., and Morgan, B.B. Effects on sustained performance of time-sharing a three-phase code transformation task (3P-Cotran). *Perceptual and Motor Skills* 33: 639–651, 1971.

Lysaght, R.J., Hill, S.G., Dick, A.O., Plamondon, B.D., Linton, P.M., Wierwille, W.W., Zaklad, A.L., Bittner, A.C., and Wherry, R.J. Operator workload: Comprehensive review and evaluation of operator workload methodologies (Technical Report 851). Alexandria, VA: Army Research Institute for the Behavioral and Social Sciences, June 1989.

2.2.16 Occlusion Secondary Task

General description – "The participant's view of a visual display is obstructed (usually by a visor). These obstructions are either initiated by the participant or imposed by the experimenter in order to determine the viewing time needed to perform a task adequately" (Lysaght et al., 1989, p. 234).

Strengths and limitations – This task can be extremely disruptive of primary-task performance.

Based on the results of four studies in which a secondary occlusion task was used, Lysaght et al. (1989) reported that performance remained stable on monitoring and driving primary tasks and degraded on driving primary tasks. Performance of the secondary occlusion task degraded when paired with primary driving tasks (see Table 2.8).

TABLE 2.8
References Listed by the Effect on Performance of Primary Tasks Paired with a Secondary Occlusion Task

	Primary Task			Secondary Task		
Type	Stable	Degraded	Enhanced	Stable	Degraded	Enhanced
Driving	Farber and Gallagher (1972)	Hicks and Wierwille (1979) Senders et al. (1967)			Farber and Gallagher (1972) Senders et al. (1967)	
Monitoring	Gould and Schaffer (1967)					

Source: From Lysaght et al. (1989, p. 250).

Data requirements – The following data are used to assess performance of this task: mean voluntary occlusion time and percent looking time/total time (Lysaght et al., 1989, p. 236).

Thresholds – Not stated.

Sources

Farber, E., and Gallagher, V. Attentional demand as a measure of the influence of visibility conditions on driving task difficulty. *Highway Research Record* 414: 1–5, 1972.

Gould, J.D., and Schaffer, A. The effects of divided attention on visual monitoring of multi-channel displays. *Human Factors* 9: 191–202, 1967.

Hicks, T.G., and Wierwille, W.W. Comparison of five mental workload assessment procedures in a moving-base driving simulator. *Human Factors* 21: 129–143, 1979.

Lysaght, R.J., Hill, S.G., Dick, A.O., Plamondon, B.D., Linton, P.M., Wierwille, W.W., Zaklad, A.L., Bittner, A.C., and Wherry, R.J. Operator workload: Comprehensive review and evaluation of operator workload methodologies (Technical Report 851). Alexandria, VA: Army Research Institute for the Behavioral and Social Sciences, June 1989.

Senders, J.W., Kristofferson, A.B., Levison, W.H., Dietrich, C.W., and Ward, J.L. The attentional demand of automobile driving. *Highway Research Record* 195: 15–33, 1967.

2.2.17 Problem-Solving Secondary Task

General description – "The participant engages in a task which requires verbal or spatial reasoning. For example, the participant might attempt to solve anagram or logic problems" (Lysaght et al., 1989, p. 233).

Human Workload

TABLE 2.9

References Listed by the Effect on Performance of Primary Tasks Paired with a Secondary Problem-Solving Task

Type	Primary Task Stable	Primary Task Degraded	Primary Task Enhanced	Secondary Task Stable	Secondary Task Degraded	Secondary Task Enhanced
Choice RT					Schouten et al. (1962)	
Driving		Merat et al. (2012) Wetherell (1981)			Merat et al. (2012) Wetherell (1981)	
Memory		Trumbo et al. (1967)			Trumbo et al. (1967)	
Monitoring	Gould and Schaffer (1967) Smith et al. (1966)					Smith et al. (1966)
Problem-Solving					Chiles and Alluisi (1979)	
Tracking		Trumbo et al. (1967)			Trumbo et al. (1967)	

Source: From Lysaght et al. (1989, p. 250).

Strengths and limitations – "This class of tasks is thought to impose heavy demands on central processing resources" (Lysaght et al., 1989, p. 233). Based on eight studies in which a problem-solving secondary task was used, Lysaght et al. (1989) reported performance remained stable on a primary monitoring task and degraded on driving, tracking, and memory primary tasks. Performance of the secondary problem-solving task remained stable when paired with a primary tracking task, degraded when paired with problem solving, driving, choice RT, and memory primary tasks, and improved when paired with a primary monitoring task (see Table 2.9).

Data requirements – The following data are used for these tasks: number of correct responses, number of incorrect responses, and mean RT for correct responses (Lysaght et al., 1989, p. 233).

Thresholds – Not stated.

Sources

Chiles, W.D., and Alluisi, E.A. On the specification of operator or occupational workload with performance-measurement methods. *Human Factors* 21: 515–528, 1979.

Gould, J.D., and Schaffer, A. The effects of divided attention on visual monitoring of multichannel displays. *Human Factors* 9: 191–202, 1967.

Lysaght, R.J., Hill, S.G., Dick, A.O., Plamondon, B.D., Linton, P.M., Wierwille, W.W., Zaklad, A.L., Bittner, A.C., and Wherry, R.J. Operator workload: Comprehensive review and evaluation of operator workload methodologies (Technical Report 851). Alexandria, VA: Army Research Institute for the Behavioral and Social Sciences, June 1989.

Merat, N., Jamson, A.H., Lai, F.C.H., and Carsten, O. Highly automated driving, secondary task performance, and driver state. *Human Factors* 54(5): 762–771, 2012.

Schouten, J.F., Kalsbeek, J.W.H., and Leopold, F.F. On the evaluation of perceptual and mental load. *Ergonomics* 5: 251–260, 1962.

Smith, R.L., Lucaccini, L.F., Groth, H., and Lyman, J. Effects of anticipatory alerting signals and a compatible secondary task on vigilance performance. *Journal of Applied Psychology* 50: 240–246, 1966.

Trumbo, D., Noble, M., and Swink, J. Secondary task interference in the performance of tracking tasks. *Journal of Experimental Psychology* 73: 232–240, 1967.

Wetherell, A. The efficacy of some auditory-vocal subsidiary tasks as measures of mental load on male and female drivers. *Ergonomics* 24: 197–214, 1981.

2.2.18 Production/Handwriting Secondary Task

General description – "The participant is required to produce spontaneous handwritten passages of prose" (Lysaght et al., 1989, p. 234).

Strengths and limitations – "With primary tasks that impose a high workload, participant's handwriting is thought to deteriorate (i.e., semantic and grammatical errors) under such conditions" (Lysaght et al., 1989, p. 234). Lysaght et al. (1989) cite a study reported by Schouten et al. (1962) in which a spontaneous writing secondary task was paired with a choice RT primary task. Performance on the secondary task degraded. A more modern version of the task is text messaging. Mouloua et al. (2010) reported a significant increase in driving errors (lane deviations, crossing the median, crashes) while text messaging than either before or after it. The data were collected in a driving simulator.

Data requirements – The number of semantic and grammatical errors is used as data for this task (Lysaght et al., 1989, p. 236).

Thresholds – Not stated.

Sources

Lysaght, R.J., Hill, S.G., Dick, A.O., Plamondon, B.D., Linton, P.M., Wierwille, W.W., Zaklad, A.L., Bittner, A.C., and Wherry, R.J. Operator workload: Comprehensive review and evaluation of operator workload methodologies (Technical Report 851). Alexandria, VA: Army Research Institute for the Behavioral and Social Sciences, June 1989.

Mouloua, M., Ahern, A., Rinalducci, E., Alberti, P., Brill, J.C., and Quevedo, A. The effects of text messaging on driver distraction: A bio-behavioral analysis. Proceedings of the Human Factors and Ergonomics Society 54th Annual Meeting, 1541–1545, 2010.

Schouten, J.F., Kalsbeek, J.W.H., and Leopold, F.F. On the evaluation of perceptual and mental load. *Ergonomics* 15: 251–260, 1962.

2.2.19 Psychomotor Secondary Task

General description – "The participant must perform a psychomotor task such as sorting different types of metal screws by size" (Lysaght et al., 1989, p. 233).

Strengths and limitations – "Tasks of this nature are thought to reflect psychomotor skills" (Lysaght et al., 1989, p. 233). Based on three studies in which a psychomotor secondary task was used, Lysaght et al. (1989) reported that performance of a tracking primary task degraded. Performance of the secondary psychomotor task degraded when paired with either a tracking or choice RT primary task (see Table 2.10). In an unusual application, Scerbo et al. (2013) reported significantly worse performance of novices than experienced surgeons on a secondary peg transfer task while performing laparoscopic surgery on a simulator.

Data requirements – The number of completed items is used to assess performance of this task.

Thresholds – Not stated.

TABLE 2.10

References Listed by the Effect on Performance of Primary Tasks Paired with a Secondary Psychomotor Task

	Primary Task			Secondary Task		
Type	Stable	Degraded	Enhanced	Stable	Degraded	Enhanced
Choice RT					Schouten et al. (1962)	
Driving					Kidd et al. (2010)[a]	
Tracking		Bergeron (1968) Wickens (1976)			Bergeron (1968)	

Source: From Lysaght et al. (1989, p. 251).
[a] Not included in Lysaght et al. (1989).

Sources

Bergeron, H.P. Pilot response in combined control tasks. *Human Factors* 10: 277–282, 1968.

Kidd, D.G., Nelson, E.T., and Baldwin, C.L. The effects of repeated exposures to collision warnings on drivers' willingness to engage in a distracting secondary task. Proceedings of the Human Factors and Ergonomics Society 54th Annual Meeting, 2086–2090, 2010.

Lysaght, R.J., Hill, S.G., Dick, A.O., Plamondon, B.D., Linton, P.M., Wierwille, W.W., Zaklad, A.L., Bittner, A.C., and Wherry, R.J. Operator workload: Comprehensive review and evaluation of operator workload methodologies (Technical Report 851). Alexandria, VA: Army Research Institute for the Behavioral and Social Sciences, June 1989.

Schouten, J.F., Kalsbeek, J.W.H., and Leopold, F.F. On the evaluation of perceptual and mental load. *Ergonomics* 5: 251–260, 1962.

Scerbo, M.W., Kennedy, R.A., Montano, M., Britt, R.C., Davis, S.S., and Stefanidis, D. A spatial secondary task for measuring laparoscopic mental workload: Differences in surgical experience. Proceedings of the Human Factors and Ergonomics Society 57th Annual Meeting, 728–732, 2013.

Wickens, C.D. The effects of divided attention on information processing in manual tracking. *Journal of Experimental Psychology: Human Perception and Performance* 2: 1–12, 1976.

2.2.20 Randomization Secondary Task

General description – "The participant must generate a random sequence of numbers, for example. It is postulated that with increased workload levels participants will generate repetitive responses (i.e., lack randomness in responses)" (Lysaght et al., 1989, p. 232).

Strengths and limitations – The task is extremely intrusive and calculating "randomness" difficult and time consuming. Based on five studies that used

TABLE 2.11

References Listed by the Effect on Performance of Primary Tasks Paired with a Secondary Randomization Task

	Primary Task			Secondary Task		
Type	Stable	Degraded	Enhanced	Stable	Degraded	Enhanced
Card Sorting	Baddeley (1966)				Baddeley (1966)	
Driving	Wetherell (1981)					
Memory		Trumbo and Noble (1970)				
Tracking	Zeitlin and Finkelman (1975)	Truijens et al. (1976)		Zeitlin and Finkelman (1975)	Truijens et al. (1976)	

Source: From Lysaght et al. (1989, p. 250).

a randomization secondary task, Lysaght et al. (1989) reported performance remained stable on tracking, card sorting, and driving primary tasks; and degraded on tracking and memory primary tasks. Performance of the secondary randomization task remained stable when paired with a tracking primary task and degraded when paired with tracking and card-sorting primary tasks (see Table 2.11).

Data requirements – The experimenter must calculate a percent redundancy score in bits of information.

Thresholds – Not stated.

Sources

Baddeley, A.D. The capacity for generating information by randomization. *Quarterly Journal of Experimental Psychology* 18: 119–130, 1966.

Lysaght, R.J., Hill, S.G., Dick, A.O., Plamondon, B.D., Linton, P.M., Wierwille, W.W., Zaklad, A.L., Bittner, A.C., and Wherry, R.J. Operator workload: Comprehensive review and evaluation of operator workload methodologies (Technical Report 851). Alexandria, VA: Army Research Institute for the Behavioral and Social Sciences, June 1989.

Truijens, C.L., Trumbo, D.A., and Wagenaar, W.A. Amphetamine and barbiturate effects on two tasks performed singly and in combination. *Acta Psychologica* 40: 233–244, 1976.

Trumbo, D., and Noble, M. Secondary task effects on serial verbal learning. *Journal of Experimental Psychology* 85: 418–424, 1970.

Wetherell, A. The efficacy of some auditory-vocal subsidiary tasks as measures of the mental load on male and female drivers. *Ergonomics* 24: 197–214, 1981.

Zeitlin, L.R., and Finkelman, J.M. Research note: Subsidiary task techniques of digit generation and digit recall indirect measures of operator loading. *Human Factors* 17: 218–220, 1975.

2.2.21 Reading Secondary Task

General description – Participants are asked to read digits or words aloud from a visual display. Measures can include: number of digits or words read, longest interval between spoken responses, longest string of consecutive digits or words, and the number of times three consecutive digits or words were spoken (Savage et al., 1978).

Strengths and limitations – This task has been sensitive to difficulty of a monitoring task. There were significant differences in the number of random digits spoken, the longest consecutive string of spoken digits, and the number of times three consecutive digits were spoken (Savage et al., 1978). The longest interval between spoken responses was not significantly different among various levels of primary task difficulty (i.e., monitoring of two, three, or four meters).

Wierwille et al. (1977) asked participants to read random digits aloud while driving in a simulator as steering ratio and wind disturbance level were manipulated. They reported that this secondary task was significantly affected by both independent variables. They concluded that it was simple to implement but may not be able to detect small changes in wind disturbance. In a follow-on study in 1978, Wierwille and Gutmann reported that this secondary task degraded primary task performance but only at low levels of workload.

Data requirements – Spoken responses must be recorded, timed, and tabulated.

Thresholds – Not stated.

Sources

Savage, R.E., Wierwille, W.W., and Cordes, R.E. Evaluating the sensitivity of various measures of operator workload using random digits as a secondary task. *Human Factors* 20(6): 649–654, 1978.

Wierwille, W.W., and Gutmann, J.C. Comparison of primary and secondary task measures as a function of simulated vehicle dynamics and driving conditions. *Human Factors* 20(2): 233–244, 1978.

Wierwille, W.W., Gutmann, J.C., Hicks, T.G., and Muto, W.H. Secondary task measurement of workload as a function of simulated vehicle dynamics and driving conditions. *Human Factors* 19(6): 557–565, 1977.

2.2.22 Simple Reaction-Time Secondary Task

General description – "The subject is presented with one discrete stimulus (either visual or auditory) and generates one response to this stimulus" (Lysaght et al., 1989, p. 232).

Strengths and limitations – This task minimizes central processing and response selection demands on the participant (Lysaght et al., 1989, p. 232).

Based on 10 studies in which a simple RT secondary task was used, Lysaght et al. (1989) reported that performance remained stable on choice RT and classification primary tasks; degraded on tracking, classification, and lexical decision tasks; and improved on detection and driving primary tasks. Performance of the secondary simple RT task degraded when paired with tracking, choice RT, memory, detection, classification, driving, and lexical decision primary tasks; and improved when paired with a tracking primary task (see Table 2.12).

Lisper et al. (1986) reported that participants who had longer RTs on driving with a simple auditory RT task on a closed course were more likely to fall asleep during an on-road drive.

Andre et al. (1995) reported significant increase in pitch, roll, and yaw error of a simulated flight task while performing a secondary simple RT task.

Data requirements – The experimenter must be able to calculate mean RT for correct responses and the number of correct responses.

Thresholds – Not stated.

Sources

Andre, A.D., Heers, S.T., and Cashion, P.A. Effects of workload preview on task scheduling during simulated instrument flight. *International Journal of Aviation Psychology* 5(1): 5–23, 1995.

Becker, C.A. Allocation of attention during visual word recognition. *Journal of Experimental Psychology: Human Perception and Performance* 2: 556–566, 1976.

Bliss, J.P., and Chancey, E. The effects of alarm system reliability and reaction training strategy on alarm systems. Proceedings of the Human Factors and Ergonomics Society 54th Annual Meeting, 2248–2252, 2010.

Comstock, E.M. Processing capacity in a letter-matching task. *Journal of Experimental Psychology* 100: 63–72, 1973.

Dodds, A.G., Clark-Carter, D., and Howarth, C.I. The effects of precueing on vibrotactile reaction times: Implications for a guidance device for blind people. *Ergonomics* 29(9): 1063–1071, 1986.

Heimstra, N.W. The effects of "stress fatigue" on performance in a simulated driving situation. *Ergonomics* 13: 209–213, 1970.

Kelly, P.A., and Klapp, S.T. Hesitation in tracking induced by a concurrent manual task. Proceedings of the 21st Annual Conference on Manual Control, 19.1–19.3, 1985.

TABLE 2.12

References Listed by the Effect on Performance of Primary Tasks Paired with a Secondary Simple RT Task

Type	Primary Task Stable	Primary Task Degraded	Primary Task Enhanced	Secondary Task Stable	Secondary Task Degraded	Secondary Task Enhanced
Choice RT	Becker (1976)				Becker (1976) Manzey et al. (2009)	
Classification	Comstock (1973)	Miller (1975)			Comstock (1973) Miller (1975)	
Detection	Bliss and Chancey (2010)[a]		Laurell and Lisper (1978)		Laurell and Lisper (1978)	
Driving			Laurell and Lisper (1978)		Laurell and Lisper (1978) Lisper et al. (1973) Libby and Chaparro (2009)	
Lexical Decision		Becker (1976)			Becker (1976)	
Memory		Dodds et al. (1986)[a]			Martin and Kelly (1974)	
Simulated Flight Task		Andre et al. (1995)[a]				
Tracking	Martin et al. (1984) for verbal response on secondary task	Heimstra (1970) Kelly and Klapp (1985) Klapp et al. (1984) Martin et al. (1984) for pointing response on secondary task Wickens and Gopher (1977)		Martin et al. (1984)	Wickens and Gopher (1977)	Heimstra (1970)

Source: From Lysaght et al. (1989, p. 251).

[a] Not included in Lysaght et al. (1989).

Klapp, S.T., Kelly, P.A., Battiste, V., and Dunbar, S. Types of tracking errors induced by concurrent secondary manual task. Proceedings of the 20th Annual Conference on Manual Control, 299–304, 1984.

Laurell, H., and Lisper, H.L. A validation of subsidiary reaction time against detection of roadside obstacles during prolonged driving. *Ergonomics* 21: 81–88, 1978.

Libby, D., and Chaparro, A. Text messaging versus talking on a cell phone: A comparison of their effects on driving performance. Proceedings of the Human Factors and Ergonomics Society 53rd Annual Meeting, 1353–1357, 2009.

Lisper, H.L., Laurell, H., and Stening, G. Effects of experience of the driver on heart-rate, respiration-rate, and subsidiary reaction time in a three-hour continuous driving task. *Ergonomics* 16: 501–506, 1973.

Lisper, H.O., Laurell, H., and van Loon, J. Relation between time to falling asleep behind the wheel on a closed course and changes in subsidiary reaction time during prolonged driving on a motorway. *Ergonomics* 29(3): 445–453, 1986.

Lysaght, R.J., Hill, S.G., Dick, A.O., Plamondon, B.D., Linton, P.M., Wierwille, W.W., Zaklad, A.L., Bittner, A.C., and Wherry, R.J. Operator workload: Comprehensive review and evaluation of operator workload methodologies (Technical Report 851). Alexandria, VA: Army Research Institute for the Behavioral and Social Sciences, June 1989.

Manzey, D., Reichenbach, J., and Onnasch, L. Human performance consequences of automated decisions aids in states of fatigue. Proceedings of the Human Factors and Ergonomics Society 53rd Annual Meeting, 329–333, 2009.

Martin, D.W., and Kelly, R.T. Secondary task performance during directed forgetting. *Journal of Experimental Psychology* 103: 1074–1079, 1974.

Martin, J., Long, J., and Broome, D. The division of attention between a primary tracing task and secondary tasks of pointing with a stylus or speaking in a simulated ship's-gunfire-control task. *Ergonomics* 27(4): 397–408, 1984.

Miller, K. Processing capacity requirements of stimulus encoding. *Acta Psychologica* 39: 393–410, 1975.

Wickens, C.D., and Gopher, D. Control theory measures of tracking as indices of attention allocation strategies. *Human Factors* 19: 349–365, 1977.

2.2.23 Simulated Flight Secondary Task

General description – "Depending on the purpose of the particular study, the subject is required to perform various maneuvers (e.g., landing approaches) under different types of conditions such as instrument flight rules or simulated crosswind conditions" (Lysaght et al., 1989, p. 234).

Strengths and limitations – This task requires extensive participant training.

Data requirements – The experimenter should record: mean error from required altitude, root-mean-square localizer error, root-mean-square glide-slope error, and number of control movements, and attitude high-pass mean square (Lysaght et al., 1989, p. 236).

Thresholds – Not stated.

Source

Lysaght, R.J., Hill, S.G., Dick, A.O., Plamondon, B.D., Linton, P.M., Wierwille, W.W., Zaklad, A.L., Bittner, A.C., and Wherry, R.J. Operator workload: Comprehensive review and evaluation of operator workload methodologies (Technical Report 851). Alexandria, VA: Army Research Institute for the Behavioral and Social Sciences, June 1989.

2.2.24 Spatial-Transformation Secondary Task

General description – "The subject must judge whether information (data) – provided by an instrument panel or radar screen – matches information which is spatially depicted by pictures or drawings of aircraft" (Lysaght et al., 1989, p. 233). "This task involves perceptual and cognitive processes" (Lysaght et al., 1989, p. 233).

Strengths and limitations – For a tracking primary task, Lysaght et al. (1989) cite work by Vidulich and Tsang (1985) in which performance of a spatial-transformation secondary task was degraded when paired with a primary tracking task. However, Kramer et al. (1984) reported a significant decrease in tracking error on a primary task when participants performed translational changes of the cursor as a secondary task.

Damos (1986) paired a visual matrix rotation task with a subtraction task. She reported no significant differences between speech responses or manual responses on the first experiment. There was a significant decrease in RTs for correct speech responses to the subtraction task when the responses were adjusted for the delay in the speech recognition system being used.

Data requirements – The following data are used to assess performance of this task: mean RT for correct responses, number of correct responses, and number of incorrect responses (Lysaght et al., 1989, p. 236).

Thresholds – Not stated.

Sources

Damos, D. The effect of using voice generation and recognition systems on the performance of dual tasks. *Ergonomics* 29(11): 1359–1370, 1986.

Kramer, A.F., Wickens, C.D., and Donchin, E. Performance enhancements under dual-task conditions. Proceedings of the 20th Annual Conference on Manual Control, 21–35, 1984.

Lysaght, R.J., Hill, S.G., Dick, A.O., Plamondon, B.D., Linton, P.M., Wierwille, W.W., Zaklad, A.L., Bittner, A.C., and Wherry, R.J. Operator workload: Comprehensive review and evaluation of operator workload methodologies (Technical Report 851). Alexandria, VA: Army Research Institute for the Behavioral and Social Sciences, June 1989.

Vidulich, M.A., and Tsang, P.S. Evaluation of two cognitive abilities tests in a dual-task environment. Proceedings of the 21st Annual Conference on Manual Control, 12.1–12.10, 1985.

2.2.25 Speed-Maintenance Secondary Task

General description – "The subject must operate a control knob to maintain a designated constant speed. This task is a psychomotor type task" (Lysaght et al., 1989, p. 234).

Strengths and limitations – This task provides a constant estimate of reserve response capacity but may be extremely intrusive on primary task performance.

Data requirements – Response is used as data for this task.

Thresholds – Not stated.

Source

Lysaght, R.J., Hill, S.G., Dick, A.O., Plamondon, B.D., Linton, P.M., Wierwille, W.W., Zaklad, A.L., Bittner, A.C., and Wherry, R.J. Operator workload: Comprehensive review and evaluation of operator workload methodologies (Technical Report 851). Alexandria, VA: Army Research Institute for the Behavioral and Social Sciences, June 1989.

2.2.26 Sternberg Memory Secondary Task

General description – The Sternberg (1966) recognition task presents a participant with a series of single letters. After each letter, the participant indicates whether that letter was or was not part of a previously memorized set of letters. RT is typically measured at two memory set sizes, two and four, and is plotted against set size (see Figure 2.2). Differences in the slope (b1 and b2 in Figure 2.2) across various design configurations indicate differences in central information processing demands. Changes in the intercept (a1 and a2 in Figure 2.2) suggest differences in either sensory or response demands. Additional data for this task include: number of correct responses and RTs for correct responses.

Strengths and limitations – The Sternberg task has been used extensively in aviation and environmental conditions assessments.

Aviation. The Sternberg task was sensitive to workload imposed by wind conditions, handling qualities, and display configurations. For example, Wolf (1978) reported the longest RTs to the Sternberg task occurred in the high gust, largest memory set (4), and poorest handling qualities condition. Schiflett (1980) reported both increased Sternberg RT and errors with degraded handling

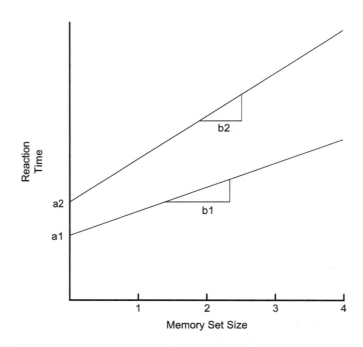

FIGURE 2.2
Sternberg memory task data.

qualities during approach and landing tasks. Similarly, Schiflett et al. (1982) reported increases in four Sternberg measures (RT for correct responses, intercept, slope, and percent errors) as handling qualities were degraded. In a helicopter application, Poston and Dunn (1986) used the Sternberg task to assess a kinesthetic tactile display. They recorded response speed and accuracy.

An advantage of the task is that it is minimally intrusive on performance of a primary flight task (Schiflett et al., 1980; Dellinger et al., 1987).

However, there are several problems associated with the use of the task. Based on data collected in a flight simulator (Taylor et al., 1985), RTs increased with workload but more so for negative than positive responses.

Further, Gawron et al. (1988) used the Sternberg memory search task to access the in-flight effects of pyridostigmine bromide. There were no significant effects of drug or crew position (pilot or copilot). This finding may be due to a true lack of a drug or crew position effect or to insensitivity of the measure. In an earlier study, Knotts and Gawron (1983) reported that the presence of the Peripheral Vision Display (PVD) reduced Sternberg RTs for one participant but not for another. They also reported that performance of the Sternberg task improved throughout the flight program and suggested extensive training before using the task in flight.

Environmental Conditions. Used alone, the Sternberg provided a reliable measure of mercury exposure (Smith and Langolf, 1981). The Sternberg has also been used as a primary task to evaluate the effects of altitude on short-term memory. Kennedy et al. (1989) reported significant increases in time per response and decreases in number correct. Rolnick and Bles (1989) also used the Sternberg but evaluated the effect of tilting and closed cabin conditions simulating ship board conditions. RT increased in the closed cabin condition compared to the no motion condition but showed no significant difference with the artificial horizon or windows conditions.

Van de Linde (1988) used a Sternberg-like test to assess the effect of wearing a respirator on the ability to recall numbers and letters from memory and to concentrate. Memory sets were 1, 2, 3, or 4 with 24 targets mixed with 120 non-targets. Letter targets took longer to find than number targets. There was no significant difference between trials 1 and 3. However, there was a 12% decrease in time while wearing the respirator.

Finally, Manzey et al. (1995) reported significant differences in dual-task performance of the Sternberg and tracking prior and during space flight. However, there was no difference in single-task performance of the Sternberg.

Effect of Dual-Task Pairing. Micalizzi and Wickens (1980) compared Sternberg RT under single and dual-task conditions. There were three manipulations of the Sternberg task: no mask, single mask, or double mask. They reported a significant increase in Sternberg RT under dual-task condition for the two masked conditions as compared to the no mask condition. Work by Micalizzi (1981) suggests that performance of the Sternberg task is poorest when it is time-shared with a failure-detection task.

Vidulich and Wickens (1986) reported that the "addition of the tracking task tends to overwhelm distinctions among the Sternberg task configurations" (p. 293). Tsang and Velazquez (1996) used the Sternberg combined with tracking to evaluate a Workload Profile measure consisting of ratings of perceptual/central, response, spatial, verbal, visual, and manual. They concluded that the variance accounted for by the memory set size variations of the Sternberg was comparable to the subjective workload ratings.

Based on the results of four studies in which a secondary Sternberg task was used, Lysaght et al. (1989) reported performance degraded on tracking, choice RT, and driving primary tasks; and improved on a tracking primary task. Performance of the secondary Sternberg task degraded when paired with a primary tracking task (see Table 2.13) or a simulated flight task.

Data requirements – The slopes and intercepts of the RT are used as data for this task. Wickens et al. (1986) have recommended that the Sternberg letters in the memorized set be changed after 50 trials. Schiflett (1983) recommended use of an adaptive interstimulus interval (ISI). Adaptation would be based on response accuracy. Knotts and Gawron (1983) recommended extensive training on the Sternberg task before data collection.

Thresholds – Not stated.

TABLE 2.13

References Listed by the Effect on Performance of Primary Tasks Paired with a Secondary Sternberg Task

	Primary Task			Secondary Task		
Type	Stable	Degraded	Enhanced	Stable	Degraded	Enhanced
Choice RT		Hart et al. (1985) Wetherell (1981) Payne et al. (1994)[a]				
Driving						
Mental Mathematics						
Simple RT		Payne et al. (1994)[a]				
Simulated Flight	Crawford et al. (1978)[a] Wierwille and Connor (1983)	O'Donnell (1976)[a]			Schiflett et al. (1982)[a] Hyman et al. (1988)	
Tracking	Tsang et al. (1996)	Wickens and Yeh (1985) Vidulich and Wickens (1986)	Briggs et al. (1972)		Briggs et al. (1972) Wickens and Yeh (1985)	

Source: From Lysaght et al. (1989, p. 252).
[a] Not included in Lysaght et al. (1989).

Sources

Briggs, G.E., Peters, G.L., and Fisher, R.P. On the locus of the divided-attention effects. *Perception and Psychophysics* 11: 315–320, 1972.

Crawford, B.M., Pearson, W.H., and Hoffman, M. *Multipurpose Digital Switching and Flight Control Workload (AMRL-TR-78-43)*. Wright-Patterson Air Force Base, OH: Air Force Aerospace Medical Research Laboratory, 1978.

Dellinger, J.A., Taylor, H.L., and Porges, S.W. Atropine sulfate effects on aviator performance and on respiratory–heart period interactions. *Aviation, Space, and Environmental Medicine* 58(4): 333–338, 1987.

Gawron, V.J., Schiflett, S., Miller, J., Ball, J., Slater, T., Parker, F., Lloyd, M., Travale, D., and Spicuzza, R.J. *The Effect of Pyridostigmine Bromide on In-Flight Aircrew Performance (USAFSAM-TR-87-24)*. Brooks AFB, TX: School of Aerospace Medicine, January 1988.

Hart, S.G., Shively, R.J., Vidulich, M.A., and Miller, R.C. The effects of stimulus modality and task integrity: Predicting dual-task performance and workload from single-task levels. Proceedings of the 21st Annual Conference on Manual Control, 5.1–5.18, 1985.

Hyman, F.C., Collins, W.E., Taylor, H. L., Domino, E.F., and Nagel, R.J. Instrument flight performance under the influence of certain combinations of antiemetic drugs. *Aviation, Space, and Environmental Medicine* 59(6): 533–539, 1988.

Kennedy, R.S., Dunlap, W.P., Banderet, L.E., Smith, M.G., and Houston, C.S. Cognitive performance deficits in a simulated climb of Mount Everest: Operation Everest II. *Aviation, Space, and Environmental Medicine* 60(2): 99–104, 1989.

Knotts, L.H., and Gawron, V.J. A preliminary flight evaluation of the peripheral vision display using the NT-33A aircraft (Report 6645-F-13). Buffalo, NY: Calspan, December 1983.

Lysaght, R.J., Hill, S.G., Dick, A.O., Plamondon, B.D., Linton, P.M., Wierwille, W.W., Zaklad, A.L., Bittner, A.C., and Wherry, R.J. Operator workload: Comprehensive review and evaluation of operator workload methodologies (Technical Report 851). Alexandria, VA: Army Research Institute for the Behavioral and Social Sciences, June 1989.

Manzey, D., Lorenz, B., Schiewe, A., Finell, G., and Thiele, G. Dual-task performance in space: Results from a single-case study during a short-term space mission. *Human Factors* 37(4): 667–681, 1995.

Micalizzi, J. The structure of processing resource demands in monitoring automatic systems (Technical Report 81-2T). Wright-Patterson AFB, OH: Air Force Institute of Technology, 1981.

Micalizzi, J., and Wickens, C.D. *The Application of Additive Factors Methodology to Workload Assessment in a Dynamic System Monitoring Task (TREPL-80-2/ONR-80-2)*. Champaign, IL: University of Illinois Engineering-Psychology Research Laboratory, December 1980.

O'Donnell, R.D. Secondary task assessment of cognitive workload in alternative cockpit configurations. In B.O. Hartman (Ed.) Higher mental functioning in operational environments (pp. C10/1–C10/5). AGARD Conference Proceedings Number 181. Neuilly sur Seine, France: Advisory Group for Aerospace Research and Development, 1976.

Payne, D.G., Peters, L.J., Birkmire, D.P., Bonto, M.A., Anatasi, J.S., and Wenger, M.J. Effects of speech intelligibility level on concurrent visual task performance. *Human Factors* 36(3): 441–475, 1994.

Poston, A.M., and Dunn, R.S. Helicopter flight evaluation of kinesthetic tactual displays: An interim report (HEL-TN-3-86). Aberdeen Proving Ground, MD: Human Engineering Laboratory, March 1986.

Rolnick, A., and Bles, W. Performance and well-being under titling conditions: The effects of visual reference and artificial horizon. *Aviation, Space, and Environmental Medicine* 60(2): 779–785, 1989.

Schiflett, S.G. Evaluation of a pilot workload assessment device to test alternate display formats and control handling qualities (SY-33R-80). Patuxent River, MD: Naval Air Test Center, July 1980.

Schiflett, S.G. Theoretical development of an adaptive secondary task to measure pilot workload for flight evaluations. Proceedings of the 27th Annual Meeting of the Human Factors Society, 602–607, 1983.

Schiflett, S., Linton, P.M., and Spicuzza, R.J. Evaluation of a pilot workload assessment device to test alternate display formats and control handling qualities. Proceedings of North Atlantic Treaty Organization (NATO) Advisory Group for Aerospace Research and Development (AGARD) (Paper Number 312). Neuilly-sur-Seine, France: AGARD, 1980.

Schiflett, S.G., Linton, P.M., and Spicuzza, R.J. Evaluation of a pilot workload assessment device to test alternative display formats and control handling qualities. Proceedings of the AIAA Workshops on Flight Testing to Identify Pilot Workload and Pilot Dynamics, 222–233, 1982.

Smith, P.J., and Langolf, G.D. The use of Sternberg's memory-scanning paradigm in assessing effects of chemical exposure. *Human Factors* 23(6): 701–708, 1981.

Sternberg, S. High speed scanning in human memory. *Science* 153: 852–654, 1966.

Taylor, H.L., Dellinger, J.A., Richardson, B.C., Weller, M.H., Porges, S.W., Wickens, C.D., LeGrand, J.E., and Davis, J.M. The effect of atropine sulfate on aviator performance (Technical Report APL-TR-85-1). Champaign, IL: University of Illinois Aviation Research Laboratory, March 1985.

Tsang, P.S., and Velazquez, V.L. Diagnosticity and multidimensional subjective workload ratings. *Ergonomics* 39(3): 358–381, 1996.

Tsang, P.S., Velaquez, V.L., and Vidulich, M.A. Viability of resource theories in explaining time-sharing performance. *Acta Psychologica* 91(2): 175–206, 1996.

Van de Linde, F.J.G. Loss of performance while wearing a respirator does not increase during a 22.5-hour wearing period. *Aviation, Space, and Environmental Medicine* 59(3): 273–277, 1988.

Vidulich, M.A., and Wickens, C.D. Causes of dissociation between subjective workload measures and performance. *Applied Ergonomics* 17(4): 291–296, 1986.

Wetherell, A. The efficacy of some auditory-vocal subsidiary tasks as measures of the mental load on male and female drivers. *Ergonomics* 24: 197–214, 1981.

Wickens, C.D., Hyman, F., Dellinger, J., Taylor, H., and Meador, M. The Sternberg memory search task as an index of pilot workload. *Ergonomics* 29: 1371–1383, 1986.

Wickens, C.D., and Yeh, Y. POCs and performance decrements: A reply to Kantowitz and Weldon. *Human Factors* 27: 549–554, 1985.

Wierwille, W.W., and Connor, S. Evaluation of 20 workload measures using a psychomotor task in a moving base aircraft simulator. *Human Factors* 25: 1–16, 1983.

Wolf, J.D. *Crew Workload Assessment: Development of a Measure of Operator Workload (AFFDL-TR-78-165).* Wright-Patterson Air Force Base, OH: Air Force Flight Dynamics Laboratory, December 1978.

2.2.27 Three-Phase Code Transformation Secondary Task

General description – "The subject operates the 3P-Cotran which is a workstation consisting of three indicator lights, a response board for subject responses, and a memory unit that the subject uses to save his/her responses. The subject must engage in a 3-phase problem-solving task by utilizing information provided by the indicator lights and recording solutions onto the memory unit" (Lysaght et al., 1989, p. 234).

Strengths and limitations – "It is a synthetic work battery used to study work behavior and sustained attention" (Lysaght et al., 1989, p. 234).

Data requirements – The following data are used to evaluate performance of this task: mean RT for different phases of response required and number of errors (resets) for different phases of response required (Lysaght et al., 1989, p. 236).

Thresholds – Not stated.

Source

Lysaght, R.J., Hill, S.G., Dick, A.O., Plamondon, B.D., Linton, P.M., Wierwille, W.W., Zaklad, A.L., Bittner, A.C., and Wherry, R.J. Operator workload: Comprehensive review and evaluation of operator workload methodologies (Technical Report 851). Alexandria, VA: Army Research Institute for the Behavioral and Social Sciences, June 1989.

2.2.28 Time-Estimation Secondary Task

General description – Participants are asked to produce a given time interval, usually 10 seconds, from the start of a stimulus, usually a tone. Measures for this task include the number of incomplete estimates and/or the length of the estimates.

Strengths and limitations – The technique has been applied in aviation display design and task performance research.

Aviation. Bortolussi et al. (1986) found a significant increase in 10-second time production intervals between easy and difficult flight scenarios. Bortolussi et al. (1987) reported similar results for a 5-second time production task.

In addition, the length of the time interval produced decreased from the beginning to the end of each flight. Gunning (1978) asked pilots to indicate when 10 seconds had passed after an auditory tone. He reported that both the number of incomplete estimates and the length of the estimates increased as workload increased. Similarly, Madero et al. (1979) reported that the number of incomplete time estimates increased over time in an aerial delivery mission. These researchers also calculated a time-estimation ratio (the length of the time estimate in flight divided by the baseline estimate). This measure was also sensitive to workload, with significant increases occurring between cruise and Initial Point (IP) and cruise and Computed Air Release Point (CARP).

Connor and Wierwille (1983) recorded time estimation mean, standard deviation, absolute error, and rmse of completed estimates during three levels of gust and aircraft stability (load). There was only one significant load effect: the standard deviation of time estimates decreased from the low to the medium load then increased from the medium to high load. This same measure was sensitive to communication load, danger, and navigation load. Specifically, Casali and Wierwille (1983) reported significant increases in time-estimation standard deviation as communication load increased. Casali and Wierwille (1984) found significant increases between low and high danger conditions. Wierwille et al. (1985b) reported the same results for navigation load, as did Hartzell (1979) for difficulty of precision hover maneuvers in a helicopter simulator.

Display Design. Bobko et al. (1986) reported, however, that verbal time estimates of a fixed time interval decreased as screen size (0.13, 0.28, and 0.58 diagonal meters) increased. In addition, men gave significantly shorter time estimates than women did.

Task Performance Research. Seidelman et al. (2012) asked 28 graduate students to move plastic beads from a dish to a bucket and from a dish to a peg. Transfer order was either alternating by color or in a pattern of four colors. During these primary tasks, participants were asked to state time when 3, 9, 15, or 21 seconds passed. Deviations from actual time were significantly longer in the four than the two-color pattern transfer as well as in the 21 second production interval than the three second production interval.

Hauser et al. (1983) reported less variability in time estimates made using a counting technique than those made without a counting technique.

Many researchers have concluded that the technique is sensitive to workload (Hart, 1978; Wierwille et al., 1985a).

Based on the results of four studies in which a time-estimation secondary task was used, Lysaght et al. (1989) reported performance on a primary flight simulation task remained stable but degraded on a primary monitoring task. Performance of the secondary time-estimation task degraded when paired with either a monitoring or a flight simulation primary task (see Table 2.14).

Data requirements – Although some researchers have reported significant differences in several time-estimation measures, the consistency of the findings using time-estimation standard deviation suggest that this may be the best time-estimation measure.

Thresholds – Not stated.

TABLE 2.14
References Listed by the Effect on Performance of Primary Tasks Paired with a Secondary Time Estimation Task

	Primary Task			Secondary Task		
Type	Stable	Degraded	Enhanced	Stable	Degraded	Enhanced
Flight Simulation	Bortolussi et al. (1987) Bortolussi et al. (1986) Casali and Wierwille (1983)[a] Kantowitz et al. (1987)[a] Wierwille and Connor (1983)[a] Wierwille et al. (1985)				Bortolussi et al. (1987, 1986) Gunning (1978)[a] Wierwille et al. (1985)	
Monitoring		Liu and Wickens (1987)			Liu and Wickens (1987)	

Source: From Lysaght et al. (1989, p. 252).
[a] Not included in Lysaght et al. (1989).

Sources

Bobko, D.J., Bobko, P., and Davis, M.A. Effect of visual display scale on duration estimates. *Human Factors* 28(2): 153–158, 1986.

Bortolussi, M.R., Hart, S.G., and Shively, R.J. Measuring moment-to-moment pilot workload using synchronous presentations of secondary tasks in a motion-base trainer. Proceedings of the Fourth Symposium on Aviation Psychology, 651–657, 1987.

Bortolussi, M.R., Kantowitz, B.H., and Hart, S.G. Measuring pilot workload in a motion base trainer: A comparison of four techniques. *Applied Ergonomics* 17: 278–283, 1986.

Casali, J.G., and Wierwille, W.W. A comparison of rating scale, secondary task, physiological, and primary task workload estimation techniques in a simulated flight emphasizing communications load. *Human Factors* 25: 623–641, 1983.

Casali, J.G., and Wierwille, W.W. On the measurement of pilot perceptual workload: A comparison of assessment techniques addressing sensitivity and intrusion issues. *Ergonomics* 27: 1033–1050, 1984.

Connor, S.A., and Wierwille, W.W. Comparative evaluation of twenty pilot workload assessment measures using a psychomotor task in a moving base aircraft simulator (Report 166457). Moffett Field, CA: NASA Ames Research Center, January 1983.

Gunning, D. Time estimation as a technique to measure workload. Proceedings of the Human Factors Society 22nd Annual Meeting, 41–45, 1978.

Hart, S.G. Subjective time estimation as an index of workload. Proceedings of the Symposium on Man-System Interface: Advances in Workload Study, 115–131, 1978.

Hartzell, E.J. Helicopter pilot performance and workload as a function of night vision symbologies. Proceedings of the 18th IEEE Conference on Decision and Control Volumes, 995–996, 1979.

Hauser, J.R., Childress, M.E., and Hart, S.G. Rating consistency and component salience in subjective workload estimation. Proceedings of the Annual Conference on Manual Control, 127–149, 1983.

Kantowitz, B.H., Bortolussi, M.R., and Hart, S.G. Measuring workload in a motion base simulation. III. Synchronous secondary task. Proceedings of the Human Factors Society 31st Annual Meeting, 834–837, 1987.

Liu, Y.Y., and Wickens, C.D. Mental workload and cognitive task automation: An evaluation of subjective and time estimation metrics (NASA 87-2). Campaign, IL: University of Illinois Aviation Research Laboratory, 1987.

Lysaght, R.J., Hill, S.G., Dick, A.O., Plamondon, B.D., Linton, P.M., Wierwille, W.W., Zaklad, A.L., Bittner, A.C., and Wherry, R.J. Operator workload: Comprehensive review and evaluation of operator workload methodologies (Technical Report 851). Alexandria, VA: Army Research Institute for the Behavioral and Social Sciences, June 1989.

Madero, R.P., Sexton, G.A., Gunning, D., and Moss, R. *Total Aircrew Workload Study for the AMST* (AFFDL-TR-79-3080, Volume 1). Wright-Patterson Air Force Base, OH: Air Force Flight Dynamics Laboratory, February 1979.

Seidelman, W., Carswell, C.M., Grant, R.C., Sublette, M., Lio, C.H., and Seales, B. Interval production as a secondary task workload measure: Consideration of primary task demands for interval selection. Proceedings of the Human Factors and Ergonomics Society 56th Annual Meeting, 1664–1668, 2012.

Wierwille, W.W., Casali, J.G., Connor, S.A., and Rahimi, M. Evaluation of the sensitivity and intrusion of mental workload estimation technique. In W. Roner (Ed.), *Advances in Man-Machine Systems Research* (vol. 2, pp. 51–127). Greenwich, CT: JAI Press, 1985a.

Wierwille, W.W., and Connor, S.A. Evaluation of 20 workload measures using a psychomotor task in a moving base aircraft simulator. *Human Factors* 25: 1–16, 1983.

Wierwille, W.W., Rahimi, M., and Casali, J.G. Evaluation of 16 measures of mental workload using a simulated flight task emphasizing mediational activity. *Human Factors* 27: 489–502, 1985b.

2.2.29 Tracking Secondary Task

General description – "The subject must follow or track a visual stimulus (target) which is either stationary or moving by means of positioning an error cursor on the stimulus using a continuous manual response device" (Lysaght et al., 1989, p. 232). Tracking tasks require nullifying an error between a desired and an actual location.

Strengths and limitations – Tracking tasks provide a continuous measure of workload. They have been used extensively in aviation. For example, several investigators (Corkindale et al., 1969; Spicuzza et al., 1974) have used a secondary tracking task to successfully evaluate workload in an aircraft simulator. Spicuzza et al. (1974) concluded that a secondary tracking task was a sensitive measure of workload. Clement (1976) used a cross-coupled tracking task in a Short Take-Off and Landing (STOL) simulator to evaluate horizontal situation displays. Andre et al. (1995) reported increased pitch, roll, and yaw rmse in a primary simulated flight task when time-shared with a secondary target-acquisition task.

The technique may be useful in ground-based simulators but inappropriate for use in flight. For example, Ramacci and Rota (1975) required flight students to perform a secondary tracking task during their initial flights. The researchers were unable to quantitatively evaluate the scores on this task due to artifacts of air turbulence and participant fatigue. Further, Williges and Wierwille (1979) state that hardware constraints make in-flight use of a secondary tracking task unfeasible and potentially unsafe.

In stationary applications, Park and Lee (1992) reported tracking task performance distinguished passing and failing groups of flight students. Manzey et al. (1995) reported significant decrements in performance of tracking and tracking with the Sternberg tasks between pre- and space-flight.

However, Damos et al. (1981) reported that dual performance of a critical tracking task improved over 15 testing sessions, suggesting a long training

TABLE 2.15
References Listed by the Effect on Performance of Primary Tasks Paired with a Secondary Tracking Task

	Primary Task				Secondary Task		
Type	Stable	Degraded	Enhanced		Stable	Degraded	Enhanced
Choice RT		Looper (1976) Whitaker (1979)				Hansen (1982) Whitaker (1979)	
Classification		Wickens et al. (1981)				Wickens et al. (1981)	
Detection		Wickens et al. (1981)				Wickens et al. (1981) Robinson and Eberts (1987)[a]	
Memory		Johnston et al. (1970)				Johnston et al. (1970)	
Monitoring	Griffiths and Boyce (1971)					Griffiths and Boyce (1971)	
Problem Solving	Wright et al. (1974)					Wright et al. (1974)	
Simple RT		Schmidt et al. (1984)				Schmidt et al. (1984)	
Simulated Flight Task		Andre et al. (1995)[a]					
Tracking	Mirchandani (1972)	Gawron (1982)[a] Hess and Teichgraber (1974) Wickens and Kessel (1980) Wickens et al. (1981)	Tsang and Wickens (1984)			Gawron (1982)[a] Tsang and Wickens (1984) Wickens et al. (1981)	Mirchandani (1972)

[a] Not included in Lysaght et al. (1989).

time to asymptote. Further, Robinson and Eberts (1987) reported degraded performance on tracking when paired with a speech warning rather than a pictorial warning. Further, Korteling (1991, 1993) reported age differences in dual-task performance of two one-dimensional compensatory tracking tasks.

Based on the results of 12 studies in which a secondary tracking task was used, Lysaght et al. (1989) reported that performance remained stable on tracking, monitoring, and problem-solving primary tasks; degraded on tracking, choice RT, memory, simple RT, detection, and classification primary tasks; and improved on a tracking primary task. Performance of the secondary tracking task degraded when paired with tracking, choice RT, memory, monitoring, problem-solving, simple RT, detection, and classification primary tasks and improved when paired with a tracking primary task (see Table 2.15).

Data requirements – The experimenter should calculate: integrated errors in mils (rmse), total time on target, total time of target, number of times of target, and number of target hits (Lysaght et al., 1989, p. 235). A secondary tracking task is most appropriate in systems when a continuous measure of workload is required. It is recommended that known forcing functions be used rather than unknown, quasi-random disturbances.

Thresholds – Not stated.

Sources

Andre, A.D., Heers, S.T., and Cashion, P.A. Effects of workload preview on task scheduling during simulated instrument flight. *International Journal of Aviation Psychology* 5(1): 5–23, 1995.

Clement, W.F. Investigating the use of a moving map display and a horizontal situation indicator in a simulated powered-lift short-haul operation. Proceedings of the 12th Annual NASA-University Conference on Manual Control, 201–224, 1976.

Corkindale, K.G.G., Cumming, F.G., and Hammerton-Fraser, A.M. Physiological assessment of a pilot's stress during landing. Proceedings of the NATO Advisory Group for Aerospace Research and Development, 56, 1969.

Damos, D., Bittner, A.C., Kennedy, R.S., and Harbeson, M.M. Effects of extended practice on dual-task tracking performance. *Human Factors* 23(5): 625–631, 1981.

Gawron, V.J. Performance effects of noise intensity, psychological set, and task type and complexity. *Human Factors* 24(2): 225–243, 1982.

Griffiths, I.D., and Boyce, P.R. Performance and thermal comfort. *Ergonomics* 14: 457–468, 1971.

Hansen, M.D. Keyboard design variables in dual-task. Proceedings of the 18th Annual Conference on Manual Control, 320–326, 1982.

Hess R.A., and Teichgraber, W.M. Error quantization effects in compensatory tracking tasks. *IEEE Transactions on Systems, Man, and Cybernetics* SMC-4: 343–349, 1974.

Johnston, W.A., Greenberg, S.N., Fisher, R.P., and Martin, D.W. Divided attention: A vehicle for monitoring memory processes. *Journal of Experimental Psychology* 83: 164–171, 1970.

Korteling, J.E. Effects of skill integration and perceptual competition on age-related differences in dual-task performance. *Human Factors* 33(1): 35–44, 1991.

Korteling, J.E. Effects of age and task similarity on dual-task performance. *Human Factors* 35(1): 1993.

Looper, M. The effect of attention loading on the inhibition of choice reaction time to visual motion by concurrent rotary motion. *Perception and Psychophysics* 20: 80–84, 1976.

Lysaght, R.J., Hill, S.G., Dick, A.O., Plamondon, B.D., Linton, P.M., Wierwille, W.W., Zaklad, A.L., Bittner, A.C., and Wherry, R.J. Operator workload: Comprehensive review and evaluation of operator workload methodologies (Technical Report 851). Alexandria, VA: Army Research Institute for the Behavioral and Social Sciences, June 1989.

Manzey, D., Lorenz, B., Schiewe, A., Finell, G., and Thiele, G. Dual-task performance in space: Results from a single-case study during a short-term space mission. *Human Factors* 37(4): 667–681, 1995.

Mirchandani, P.B. An auditory display in a dual axis tracking task. *IEEE Transactions on Systems, Man, and Cybernetics* 2: 375–380, 1972.

Park, K.S., and Lee, S.W. A computer-aided aptitude test for predicting flight performance of trainees. *Human Factors* 34(2): 189–204, 1992.

Ramacci, C.A., and Rota, P. Flight fitness and psycho-physiological behavior of applicant pilots in the first flight missions. Proceedings of NATO Advisory Group for Aerospace Research and Development (N7B-24304), vol. 153, B8, 1975.

Robinson, C.P., and Eberts, R.E. Comparison of speech and pictorial displays in a cockpit environment. *Human Factors* 29(1): 31–44, 1987.

Schmidt, K.H., Kleinbeck, U., and Brockman, W. Motivational control of motor performance by goal setting in a dual-task situation. *Psychological Research* 46: 129–141, 1984.

Spicuzza, R.J., Pinkus, A.R., and O'Donnell, R.D. Development of performance assessment methodology for the digital avionics information system. Dayton, OH: Systems Research Laboratories, August 1974.

Tsang, P.S., and Wickens, C.D. The effects of task structures on time-sharing efficiency and resource allocation optimality. Proceedings of the 20th Annual Conference on Manual Control, 305–317, 1984.

Whitaker, L.A. Dual task interference as a function of cognitive processing load. *Acta Psychologica* 43: 71–84, 1979.

Wickens, C.D., and Kessel, C. Processing resource demands of failure detection in dynamic systems. *Journal of Experimental Psychology: Human Perception and Performance* 6: 564–577, 1980.

Wickens, C.D., Mountford, S.J., and Schreiner, W. Multiple resources, task-hemispheric integrity, and individual differences in time-sharing. *Human Factors* 23: 211–229, 1981.

Williges, R.C., and Wierwille, W.W. Behavioral measures of aircrew mental workload. *Human Factors* 21: 549–574, 1979.

Wright, P., Holloway, C.M., and Aldrich, A.R. Attending to visual or auditory verbal information while performing other concurrent tasks. *Quarterly Journal of Experimental Psychology* 26: 454–463, 1974.

2.2.30 Workload Scale Secondary Task

General description – A workload scale was developed by tallying the number of persons who performed better in each task combination of the Multiple Task Performance Battery (MTPB) and converting these proportions to z scores (Chiles and Alluisi, 1979). The resulting z scores are multiplied by −1 so that the most negative score is associated with the lowest workload.

Strengths and limitations – Workload scales are easy to calculate but have two assumptions: (1) linear additivity and (2) no interaction between tasks. Some task combinations may violate these assumptions. Further, the intrusiveness of secondary tasks may preclude their use in nonlaboratory settings.

Data requirements – Performance of multiple combinations of tasks is required.

Thresholds – Dependent on task and task combinations being used.

Source

Chiles, W.D., and Alluisi, E.A. On the specification of operator or occupational workload with performance-measurement methods. *Human Factors* 21(5): 515–528, 1979.

2.3 Subjective Measures of Workload

There are five types of subjective measures of workload. The first is comparison measures in which the participant is asked which of two tasks has the higher workload. This type of measure is described in Section 2.3.1. The second type is decision tree in which the participant is stepped through a series of discrete questions to reach a single workload rating (see Section 2.3.2). The third type of subjective workload measure is a set of subscales, each of which was designed to measure different aspects of workload (see Section 2.3.3). The fourth type is single number, which as the name implies requires the participant to give only one number to rate the workload (Section 2.3.4). The final type of subjective measure of workload is task-analysis based. These measures break the tasks into subtasks and subtask requirements for workload evaluation (see Section 2.3.5). A summary is provided in Table 2.16.

TABLE 2.16
Comparison of Subjective Measures of Workload

Section	Measure	Reliability	Task Time	Ease of Scoring
2.3.4.1	Air Traffic Workload Input Technique	High	Requires rating 1 to 7	No scoring needed
2.3.1.1	Analytical Hierarchy Process	High	Requires rating pairs of tasks	Computer scored
2.3.5.1	Arbeitswissenshaftliches Erhebungsverfahren zur Tatigkeitsanalyze	High	Requires rating 216 items	Requires multivariate statistics
2.3.3.1	Assessing the Impact of Automation on Mental Workload	High	Requires rating 32 items	Requires calculating percentage
2.3.2.1	Bedford Workload Scale	High	Requires two decisions	No scoring needed
2.3.5.2	Computerized Rapid Analysis of Workload	Unknown	None	Requires detailed mission timeline
2.3.4.2	Continuous Subjective Assessment of Workload	High	Requires programming computer prompts	Computer scored
2.3.2.2	Cooper-Harper Rating Scale	High	Requires three decisions	No scoring needed
2.3.4.2	Continuous Subjective Assessment of Workload	Unknown	Rating 1 to 10 while viewing video of flight	No scoring needed
2.3.3.2	Crew Status Survey	High	Requires one decision	No scoring needed
2.3.4.3	Dynamic Workload Scale	High	Requires ratings by pilot and observer whenever workload changes	No scoring needed
2.3.4.4	Equal-Appearing Intervals	Unknown	Requires ratings in several categories	No scoring needed
2.3.3.3	Finegold Workload Rating Scale	High	Requires five ratings	Requires calculating an average
2.3.3.4	Flight Workload Questionnaire	May evoke response bias	Requires four ratings	No scoring needed
2.3.4.5	Hart and Bortolussi Rating Scale	Unknown	Requires one rating	No scoring needed

(Continued)

TABLE 2.16 (CONTINUED)

Comparison of Subjective Measures of Workload

Section	Measure	Reliability	Task Time	Ease of Scoring
2.3.3.5	Hart and Hauser Rating Scale	Unknown	Requires six ratings	Requires interpolating quantity from mark on scale
2.3.2.3	Honeywell Cooper-Harper Rating Scale	Unknown	Requires three decisions	No scoring needed
2.3.3.6	Human Robot Interaction Workload Measurement Tool	Unknown	Requires rating 1 to 5 on six items	Post rating interview
2.3.4.6	Instantaneous Self Assessment (ISA)	High	Requires rating of 1 to 5	No scoring needed
2.3.1.2	Magnitude Estimation	Moderate	Requires comparison to a standard	No scoring needed
2.3.5.3	McCracken-Aldrich Technique	Unknown	May require months of preparation	Requires computer programmer
2.3.4.7	McDonnell Rating Scale	Unknown	Requires three or four decisions	No scoring needed
2.3.2.4	Mission Operability Assessment Technique	Unknown	Requires two ratings	Requires conjoint measurement techniques
2.3.2.5	Modified Cooper-Harper Rating Scale	High	Requires three decisions	No scoring needed
2.3.3.7	Multi-Descriptor Scale	Low	Requires six ratings	Requires calculating an average
2.3.3.8	Multidimensional Rating Scale	High	Requires eight ratings	Requires measurement of line length
2.3.3.9	Multiple Resources Questionnaire	Moderate	Requires 16 ratings	No scoring needed
2.3.3.10	NASA Bipolar Rating Scale	High	Requires 10 ratings	Requires weighting procedure
2.3.3.11	NASA Task Load Index	High	Requires six ratings	Requires weighting procedure
2.3.4.8	Overall Workload Scale	Moderate	Requires one rating	No scoring needed
2.3.4.9	Pilot Objective/Subjective Workload Assessment Technique	High	Requires one rating	No scoring needed

(Continued)

TABLE 2.16 (CONTINUED)

Comparison of Subjective Measures of Workload

Section	Measure	Reliability	Task Time	Ease of Scoring
2.3.1.3	Pilot Subjective Evaluation	Unknown	Requires rating systems on four scales and completion of questionnaire	Requires extensive interpretation
2.3.3.12	Profile of Mood States	High	Requires about 10 minutes to complete	Requires manual or computer scoring
2.3.2.6	Sequential Judgment Scale	High	Requires rating each task	Requires measurement and conversion to percent
2.3.3.13	Subjective Workload Assessment Technique	High	Requires prior card sort and three ratings	Requires computer scoring
2.3.1.4	Subjective Workload Dominance Technique	High	Requires N(N-1)/2 paired comparisons	Requires calculating geometric means
2.3.5.4	Task Analysis Workload	Unknown	May require months of preparation	Requires detailed task analysis
2.3.3.14	Team Workload Questionnaire	Unknown	Requires rating 10 items from 1 to 10	Divide by 10 to get score 1 to 100
2.3.4.10	Utilization	High	None	Requires regression
2.3.3.15	Workload/Compensation/ Interference/Technical Effectiveness	Unknown	Requires ranking 16 matrix cells	Requires complex mathematical processing
2.3.5.5	Zachary/Zaklad Cognitive Analysis	Unknown	May require months of preparation	Requires detailed task analysis

Casali and Wierwille (1983) identified several advantages of subjective measures: "inexpensive, unobtrusive, easily administered, and readily transferable to full-scale aircraft and to a wide range of tasks" (p. 640). Gopher (1983) concluded that subjective measures "are well worth the bother" (p. 19). Wickens (1984) states that subjective measures have high face validity. Muckler and Seven (1992) state that subjective measures may be essential.

O'Donnell and Eggemeier (1986), however, identified six limitations of subjective measures of workload: (1) potential confound of mental and physical workload, (2) difficulty in distinguishing external demand/task difficulty from actual workload, (3) unconscious processing of information that the operator cannot rate subjectively, (4) dissociation of subjective ratings and task performance, (5) require well-defined question, and (6) dependence on short-term memory. Eggemeier (1981) identified two additional issues: (1) developing a generalized measure of subjective mental workload and (2) identified factors related to the subjective experience of workload.

In addition, Meshkati et al. (1990) warn that raters may interpret the words in a rating scale differently, thus leading to inconsistent results. Finally, Heiden and Caldwell (2015) discussed use of subjective mental workload measures in populations with cognitive deficits such as Traumatic Brain Injury (TBI).

Sources

Casali, J.G., and Wierwille, W.W. A comparison of rating scale, secondary task, physiological, and primary-task workload estimation techniques in a simulated flight task emphasizing communications load. *Human Factors* 25: 623–642, 1983.

Eggemeier, F.T. Current issues in subjective assessment of workload. Proceedings of the Human Factors Society 25th Annual Meeting, 513–517, 1981.

Gopher, D. *The Workload Book: Assessment of Operator Workload to Engineering Systems (NASA-CR-166596)*. Moffett Field, CA: NASA Ames Research Center, November 1983.

Heiden, S.M., and Caldwell, B.S. Considerations for using subjective mental workload measures in populations with cognitive deficits. Proceedings of the 59th Human Factors and Ergonomics Society Annual Meeting, 476–480, 2015.

Meshkati, N., Hancock, P.A., and Rahimi, M. Techniques in mental workload assessment. In J.R. Wilson and E.N. Corlett (Eds.) *Evaluation of Human Work. A Practical Ergonomics Methodology* (pp. 605–627). New York: Taylor & Francis Group, 1990.

Muckler, F.A., and Seven, S.A. Selecting performance measures "objective" versus "subjective" measurement. *Human Factors* 34: 441–455, 1992.

O'Donnell, R.D., and Eggemeier, F.T. Workload assessment methodology. In K.R. Boff, L. Kaufman, and J.P. Thomas (Eds.) *Handbook of Perception and Human Performance* (pp. 42-1–42-89). New York: Wiley and Sons, 1986.

Wickens, C.D. *Engineering Psychology and Human Performance*. Columbus, OH: Charles E. Merrill, 1984.

2.3.1 Comparison of Subjective Workload Measures

Comparison measures require the participant to identify which of two tasks has the higher workload. Examples include the Analytical Hierarchy Process (Section 2.3.1.1), Magnitude Estimation (Section 2.3.1.2), Pilot Subjective Evaluation (Section 2.3.1.3), and Subjective Workload Dominance (Section 2.3.1.4).

2.3.1.1 Analytical Hierarchy Process

General description – The analytical hierarchy process (AHP) uses the method of paired comparisons to measure workload. Specifically, participants rate which of a pair of conditions has the higher workload. All combinations of conditions must be compared. Therefore, if there are n conditions, the number of comparisons is 0.5n(n-1).

Strengths and limitations – Lidderdale (1987) found high consensus in the ratings of both pilots and navigators for a low-level tactical mission. Vidulich and Tsang (1987) concluded that AHP ratings were more valid and reliable than either an overall workload rating or NASA-TLX. Vidulich and Bortolussi (1988) reported that AHP ratings were more sensitive to attention than secondary RTs. Vidulich and Tsang (1988) reported high test/retest reliability. Bortolussi and Vidulich (1991) reported significantly higher workload using speech controls than manual controls in a combat helicopter simulated mission. AHP accounted for 64.2% of the variance in the mission phase (Vidulich and Bortolussi, 1988). It has been used in China to determine worker salaries based on "job intensity" derived from physical and mental loading, environmental conditions, and danger (Shen et al., 1990).

AHP was also sensitive to degree of automation in a combat helicopter simulation (Bortolussi and Vidulich, 1989). Metta (1993) used the AHP to develop a rank ordering of computer interfaces. She identified the following advantages of AHP: (1) easy to quantify consistency in human judgments, (2) yields useful results in spite of small sample sizes and low probability of statistically significant results, and (3) requires no statistical assumptions. However, complex mathematical procedures must be employed (Lidderdale, 1987; Lidderdale and King, 1985; Saaty, 1980).

Data requirements – Four steps are required to use the AHP. First, a set of instructions must be written. A verbal review of the instructions should be conducted after the participants have read the instructions to ensure their understanding of the task. Second, a set of evaluation sheets must be designed to collect the participants' data. An example is presented in Figure 2.3. Each sheet has the two conditions to be compared in separate columns, one on the right side of the page, the other on the left. A 17-point rating scale is placed between the two sets of conditions. The scale uses five descriptors in a predefined order and allows a single point between each for mixed ratings. Vidulich (1988) defined the scale descriptors (see Table 2.17). Budescu et al. (1986) provide critical value tables for detecting inconsistent judgments and participants.

Third, the data must be scored. The scores range from +8 (absolute dominance of the left-side condition over the right-side condition) to −8 (absolute dominance of the right-side condition over the left-side condition). Finally,

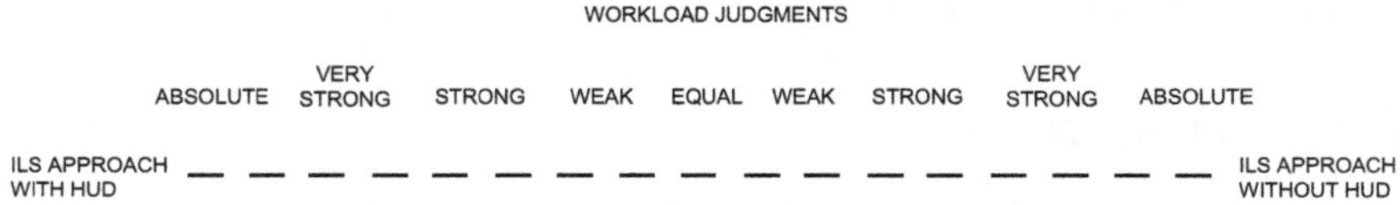

WORKLOAD JUDGMENTS

ABSOLUTE　VERY STRONG　STRONG　WEAK　EQUAL　WEAK　STRONG　VERY STRONG　ABSOLUTE

ILS APPROACH WITH HUD — — — — — — — — — — — — — — — — — ILS APPROACH WITHOUT HUD

FIGURE 2.3
Example AHP rating scale.

TABLE 2.17

Definitions of AHP Scale Descriptors

EQUAL	The two task combinations are absolutely equal in the amount of workload generated by the simultaneous tasks.
WEAK	Experience and judgment slightly suggest that one of the combinations of tasks has more workload than the other.
STRONG	Experience and judgment strongly suggest that one of the combinations has higher workload.
VERY STRONG	One task combinations is strongly dominant in the amount of workload, and this dominance is clearly demonstrated in practice.
ABSOLUTE	The evidence supporting the workload dominance of one task combination is the highest possible order of affirmation (adapted from Vidulich, 1988, p. 5).

the scores are input, in matrix form, into a computer program. The output of this program is a scale weight for each condition and three measures of goodness of fit.

Thresholds – Not stated.

Sources

Bortolussi, M.R., and Vidulich, M.A. The effects of speech controls on performance in advanced helicopters in a double stimulation paradigm. Proceedings of the International Symposium on Aviation Psychology, 216–221, 1991.

Bortolussi, M.R., and Vidulich, M.A. The benefits and costs of automation in advanced helicopters: An empirical study. Proceedings of the 5th International Symposium on Aviation Psychology, 594–599, 1989.

Budescu, D.V., Zwick, R., and Rapoport, A. A comparison of the Eigen value method and the geometric mean procedure for ratio scaling. *Applied Psychological Measurement* 10: 68–78, 1986.

Lidderdale, I.G. Measurement of aircrew workload during low-level flight, practical assessment of pilot workload (AGARD-AG-282). Proceedings of NATO Advisory Group for Aerospace Research and Development (AGARD), 69–77, 1987.

Lidderdale, I.G., and King, A.H. *Analysis of Subjective Ratings Using the Analytical Hierarchy Process: A Microcomputer Program.* High Wycombe, England: OR Branch NFR, HQ ST C, RAF, 1985.

Metta, D.A. An application of the analytic hierarchy process: A rank-ordering of computer interfaces. *Human Factors* 35(1): 141–157, 1993.

Saaty, T.L. The Analytical Hierarchy Process. New York: McGraw-Hill, 1980.

Shen, R., Meng, X., and Yan, Y. Analytic hierarchy process applied to synthetically evaluate the labor intensity of jobs. *Ergonomics* 33(7): 867, 1990.

Vidulich, M.A. Notes on the AHP procedure, 1988. (Available from Dr. Michael A. Vidulich, Wright-Patterson Air Force Base, OH 45433-6573.)

Vidulich, M.A., and Bortolussi, M.R. A dissociation of objective and subjective workload measures in assessing the impact of speech controls in advanced helicopters. Proceedings of the Human Factors Society 32nd Annual Meeting, 1471–1475, 1988.

Vidulich, M.A., and Tsang, P.S. Absolute magnitude estimation and relative judgment approaches to subjective workload assessment. Proceedings of the Human Factors Society 31st Annual Meeting, 1057–1061, 1987.

Vidulich, M.A., and Tsang, P.S. Evaluating immediacy and redundancy in subjective workload techniques. Proceedings of the Twenty-Third Annual Conference on Manual Control, 1988.

2.3.1.2 Magnitude Estimation

General description – Participants are required to estimate workload numerically in relation to a standard.

Strengths and limitations – Borg (1978) successfully used this method for evaluating workload. Helm and Heimstra (1981) reported a high correlation between workload estimates and task difficulty. Masline (1986) reported sensitivity comparable to estimates from the equal-appearing intervals method and SWAT. Kramer et al. (1987) reported good correspondence to performance in a fixed-based flight simulator. In contrast, Gopher and Braune (1984) found a low correlation between workload estimates and reaction-time performance. Hart and Staveland (1988) suggested that the presence of a standard enhances inter-rater reliability. O'Donnell and Eggemeier (1986), however, warned that participants may be unable to retain an accurate memory of the standard over the course of an experiment.

Data requirements – A standard must be well defined.

Thresholds – Not stated.

Sources

Borg, C.G. Subjective aspects of physical and mental load. *Ergonomics* 21: 215–220, 1978.

Gopher, D., and Braune, R. On the psychophysics of workload: Why bother with subjective measures? *Human Factors* 26: 519–532, 1984.

Hart, S.G., and Staveland, L.E. Development of NASA-TLX (Task Load Index): Results of empirical and theoretical research. In P.A. Hancock and N. Meshkati (Eds.) *Human Mental Workload*. Amsterdam: Elsevier, 1988.

Helm, W., and Heimstra, N.W. The relative efficiency of psychometric measures of task difficulty and task performance in predictive task performance (Report No. HFL-81-5). Vermillion, SD: University of South Dakota, Psychology Department, Human Factors Laboratory, 1981.

Human Workload

Kramer, A.F., Sirevaag, E.J., and Braune, R. A psychophysical assessment of operator workload during simulated flight missions. *Human Factors* 29: 145–160, 1987.

Masline, P.J. A comparison of the sensitivity of interval scale psychometric techniques in the assessment of subjective workload. Unpublished master's thesis, University of Dayton, Dayton, OH, 1986.

O'Donnell, R.D., and Eggemeier, F.T. Workload assessment methodology. In K.R. Boff, L. Kaufman, and J. Thomas (Eds.) *Handbook of Perception and Human Performance. Vol. 2, Cognitive Processes and Performance* (pp. 1–49). New York: Wiley, 1986.

2.3.1.3 Pilot Subjective Evaluation

General description – The Pilot Subjective Evaluation (PSE) workload scale (see Figure 2.4) was developed by Boeing for use in the certification of the Boeing 767 aircraft. The scale is accompanied by a questionnaire. Both the scale and the questionnaire are completed with reference to an existing aircraft selected by the pilot.

Strengths and limitations – Fadden (1982) and Ruggerio and Fadden (1987) stated that the ratings of workload greater than the reference aircraft were useful in identifying aircraft design deficiencies.

Data requirements – Each participant must complete both the PSE scale and the questionnaire.

Thresholds – 1, minimum workload; 7, maximum workload.

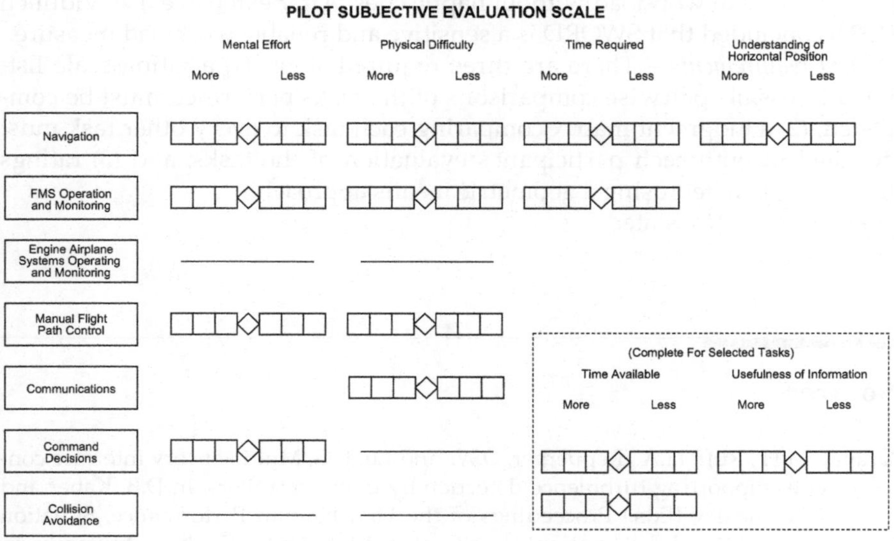

FIGURE 2.4
Pilot subjective evaluation scale (from Lysaght et al., 1989, p. 107).

Sources

Fadden, D. Boeing Model 767 flight deck workload assessment methodology. Presented at the SAE Guidance and Control System Meeting, Williamsburg, VA, 1982.

Lysaght, R.J., Hill, S.G., Dick, A.O., Plamondon, B.D., Linton, P.M., Wierwille, W.W., Zaklad, A.L., Bittner, A.C., and Wherry, R.J. Operator workload: Comprehensive review and evaluation of operator workload methodologies (Technical Report 851). Alexandria, VA: Army Research Institute for the Behavioral and Social Sciences, June 1989.

Ruggerio, F., and Fadden, D. Pilot subjective evaluation of workload during a flight test certification programme. In A.H. Roscoe (Ed.) *The Practical Assessment of Pilot Workload.* AGARD-oegraph 282 (pp. 32–36). Neuilly-sur-Seine, France: AGARD, 1987.

2.3.1.4 Subjective Workload Dominance

General description – The Subjective Workload Dominance (SWORD) technique uses judgment matrices to assess workload.

Strengths and limitations – SWORD has been useful in projecting workload associated with various Head Up Display (HUD) formats (Vidulich et al., 1991). In addition, Tsang and Vidulich (1994) reported significant differences in SWORD ratings as a function of tracking task condition. The test-retest reliability was +0.937. In a UAV study, Draper et al. (2000) reported a significant increase in workload without haptic cues. After extensive use, Vidulich (1989) concluded that SWORD is a sensitive and reliable workload measure.

Data requirements – There are three required steps: (1) a rating scale listing all possible pairwise comparisons of the tasks performed must be completed, (2) a judgment matrix comparing each task to every other task must be filled in with each participant's evaluation of the tasks, and (3) ratings must be calculated using a geometric means approach.

Thresholds – Not stated.

Sources

Draper, M.H., Ruff, H.A., Repperger, D.W., and Lu, L.G. Multi-sensory interface concepts supporting turbulence detection by UAV controllers. In D.B. Kaber and M.R. Endsley (Eds.) Proceedings of the First Human Performance, Situation Awareness and Automation: User-Centered Design for the New Millennium, 107–112, 2000.

Tsang, P.S., and Vidulich, M.A. The roles of immediacy and redundancy in relative subjective workload assessment. *Human Factors* 36(3): 503–513, 1994.

Vidulich, M.A. The use of judgment matrices in subjective workload assessment: The subjective workload dominance (SWORD) technique. Proceedings of the Human Factors Society 33rd Annual Meeting, 1406–1410, 1989.

Vidulich, M.A., Ward, G.F., and Schueren, J. Using the Subjective Workload Dominance (SWORD) technique for projective workload assessment. *Human Factors* 33(6): 677–691, 1991.

2.3.2 Decision Tree Subjective Workload Measures

Decision tree subjective measures of workload require the participant to step through a series of discrete questions to reach a single workload rating. Examples include: the Bedford Workload Scale (Section 2.3.2.1), Cooper-Harper Rating Scale (Section 2.3.2.2), Honeywell Copper-Harper Rating Scale (Section 2.3.2.3), Mission Operability Assessment Technique (Section 2.3.2.4), Modified Cooper-Harper Rating Scale (Section 2.3.2.5), and Sequential Judgment Scale (Section 2.3.2.6)

2.3.2.1 Bedford Workload Scale

General description – Roscoe (1984) described a modification of the Cooper-Harper Rating Scale created by trial and error with the help of test pilots at the Royal Aircraft Establishment at Bedford, England. The Bedford Workload Scale (see Figure 2.5) retained the binary decision tree and the four- and 10-rank ordinal structures of the Cooper-Harper Rating Scale. The three-rank ordinal structure asked pilots to assess whether: (1) it was possible to complete the task, (2) the workload was tolerable, and (3) the workload was satisfactory without reduction. The rating-scale end points were: *workload insignificant* to *task abandoned*. In addition to the structure, the Cooper-Harper (1969) definition of pilot workload was used: "the integrated mental and physical effort required to satisfy the perceived demands of a specified flight task" (Roscoe, 1984, p. 12–8). The concept of spare capacity was used to help define levels of workload.

Strengths and limitations – Roscoe (1987) reported that the scale was well accepted by aircrews. Roscoe (1984) reported that pilots found the scale "easy to use without the need to always refer to the decision tree." He also noted that it was necessary to accept ratings of 3.5 from the pilots. These statements suggest that the pilots emphasized the 10- rather than the four-rank, ordinal structure of the Bedford Workload Scale. Roscoe (1984) found that pilot workload ratings and heart rates varied in similar manners during close-coupled in-flight maneuvers in a BAE 125 twinjet aircraft. He felt that the heart-rate information complemented and increased the value of subjective workload ratings. He also noted the lack of absolute workload information provided by the Bedford Workload Scale and by heart-rate data.

Wainwright (1987) used the scale during certification of the BAE 146 aircraft. Burke et al. (2016) used the Bedford Scale to measure the workload of

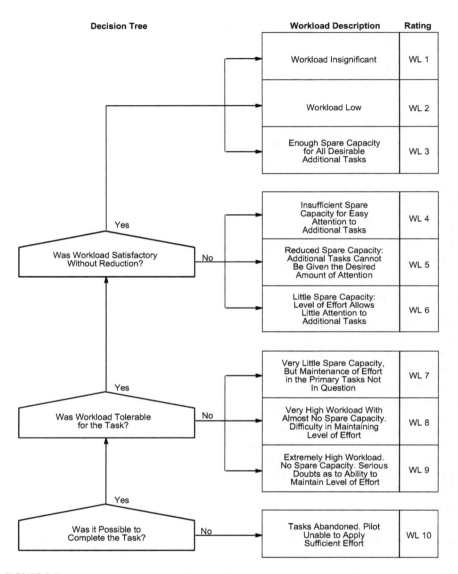

FIGURE 2.5
Bedford Workload Scale.

pilots using a system to identify trajectory improvements over the current flight path. The data were collected in flight from 12 pilots. Corwin et al. (1989) concluded that the Bedford Workload Scale is a reliable and valid measure of workload based on flight simulator data.

Svensson et al. (1997) reported the reliability of the Bedford Workload Scale to be +0.82 among 18 pilots flying simulated low-level, high-speed missions. The correlation with the NASA TLX was +0.826 and with the SWAT was +0.687. Tsang and Johnson (1987) concluded that the Bedford Workload Scale provided a good measure of spare capacity.

However, Oman et al. (2001) reported no increase in Bedford Workload Scale ratings between different vertical display formats for fixed-wing aircraft. There were reductions in root mean square altitude scores, however. Further, Comstock et al. (2010) reported no significant differences in workload between pilot flying and pilot monitoring during simulated surface/taxi operations using four types of communication (voice/paper, data/paper, data/moving map, and data/moving map route). In addition, Vidulich and Bortolussi (1988) reported significant differences in Bedford Workload Scale ratings across four flight segments. However, the workload during hover was rated less than that during hover with a simultaneous communication task. Further, the scale was not sensitive to differences in either control configurations nor combat countermeasure conditions.

Vidulich (1991) questioned whether the scale measures space capacity. In addition, Lidderdale (1987) reported that post-flight ratings were very difficult for aircrews to make. Finally, Ed George (personal communication, 2002) analyzed the responses to a survey examining the terminology used in the Bedford Workload Scale. Of the 20 United States (U.S.) Air Force pilots, four reported confusion between WL1 and WL2, two between WL2 and WL3, 10 between WL4 and WL5, six between WL5 and WL6, two between WL6 and WL7, and three between WL8 and WL9.

Data requirements – Roscoe (1984) suggested the use of short, well-defined flight tasks to enhance the reliability of subjective workload ratings. Harris et al. (1992) state that "some practice" is necessary to become familiar with the scale. They also suggest the use of non-parametric analysis technique since the Bedford Workload Scale is not an interval scale.

Thresholds – Minimum value is 1, maximum is 10. Sturrock and Fairburn (2005) defined two sets of red line values for the Bedford Rating Scale:

Development/risk reduction workload assessments

1, 2, 3 acceptable

4, 5, 6, 7, 8 investigate further

9–10 unacceptable, probable design change

Qualification workload assessments

1–8 acceptable

9–10 investigate design change (p. 590)

Sources

Burke, K.A., Wing, D.J., and Haynes, M. Flight test assessments of pilot workload, system usability, and situation awareness of TASAR. Proceedings of the Human Factors and Ergonomics Society 60th Annual Meeting, 61–65, 2016.

Comstock, J.R., Baxley, B.T., Norman, R.M., Ellis, K.K.E., Adams, C.A., Latorella, K.A., and Lynn, W.A. The impact of data communication messages in the terminal area on flight crew workload and eye scanning. Proceedings of the Human Factors and Ergonomics Society 54th Annual Meeting, 121–125, 2010.

Corwin, W.H., Sandry-Garza, D.L., Biferno, M.H., Boucek, G.P., Logan, A.L., Jonsson, J.E., and Metalis, S.A. *Assessment of Crew Workload Measurement Methods, Techniques and Procedures. Volume I – Process, Methods, and Results (WRDC-TR-89-7006)*. Wright-Patterson Air Force Base, OH: Wright Research and Development Center, 1989.

Harris, R.M., Hill, S.G., Lysaght, R.J., and Christ, R.E. *Handbook for Operating the OWL & NEST Technology (ARI Research Note 92-49)*. Alexandria, VA: United States Army Research Institute for the Behavioral and Social Sciences, 1992.

Lidderdale, I.G. Measurement of aircrew workload during low-level flight, practical assessment of pilot workload (AGARD-AG-282). Proceedings of NATO Advisory Group for Aerospace Research and Development (AGARD). Neuilly-sur-Seine, France: AGARD, 1987.

Oman, C.M., Kendra, A.J., Hayashi, M., Stearns, M.J., and Burki-Cohen, J. Vertical navigation displays: Pilot performance and workload during simulated constant angle of descent GPS approaches. *International Journal of Aviation Psychology* 11 (1): 15–31, 2001.

Roscoe, A.H. Assessing pilot workload in flight. Flight test techniques. Proceedings of NATO Advisory Group for Aerospace Research and Development (AGARD) (AGARD-CP-373). Neuilly-sur-Seine, France: AGARD, 1984.

Roscoe, A.H. In-flight assessment of workload using pilot ratings and heart rate. In A.H. Roscoe (Ed.) *The Practical Assessment of Pilot Workload*. AGARDograph No. 282 (pp. 78–82). Neuilly-sur-Seine, France: AGARD, 1987.

Sturrock, F., and Fairburn, C. Measuring pilot workload in single and multi-crew aircraft. Measuring pilot workload in a single and multi-crew aircraft. Contemporary Ergonomics 2005: Proceedings of the International Conference on Contemporary Ergonomics (CE2005), 588–592, 2005.

Svensson, E., Angelborg-Thanderz, M., Sjoberg, L., and Olsson, S. Information complexity – Mental workload and performance in combat aircraft. *Ergonomics* 40(3): 362–380, 1997.

Tsang, P.S., and Johnson, W. Automation: Changes in cognitive demands and mental workload. Proceedings of the 4th Symposium on Aviation Psychology, 616–622, 1987.

Vidulich, M.A. The Bedford Scale: Does it measure spare capacity? Proceedings of the 6th International Symposium on Aviation Psychology, vol. 2, 1136–1141, 1991.

Vidulich, M.A., and Bortolussi, M.R. Control configuration study. Proceedings of the American Helicopter Society National Specialist's Meeting: Automation Applications for Rotorcraft, 20–29, 1988.

Wainwright, W. Flight test evaluation of crew workload. In A.H. Roscoe (Ed.) *The Practical Assessment of Pilot Workload*. AGARDograph No. 282 (pp. 60–68). Neuilly-sur-Seine, France: AGARD, 1987.

2.3.2.2 Cooper-Harper Rating Scale

General description – The Cooper-Harper Rating Scale is a decision tree that uses adequacy for the task, aircraft characteristics, and demands on the pilot to rate handling qualities of an aircraft (see Figure 2.6).

Human Workload

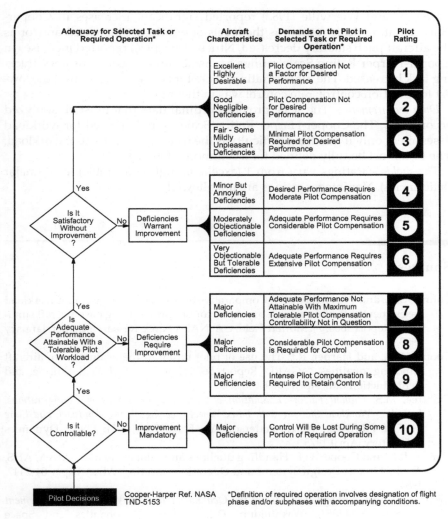

FIGURE 2.6
Cooper-Harper Rating Scale.

Strengths and limitations – The Cooper-Harper Rating Scale is the current standard for evaluating aircraft handling qualities. It reflects differences in both performance and workload and is behaviorally anchored. It requires minimum training and a briefing guide has been developed (see Cooper and Harper, 1969, pp. 34–39). Harper and Cooper (1984) describe a series of evaluations of the rating scale.

Cooper-Harper ratings have been sensitive to variations in controls, displays, and aircraft stability (Crabtree, 1975; Krebs and Wingert, 1976; Lebacqz and Aiken, 1975; Schultz et al., 1970; Wierwille and Connor, 1983).

Connor and Wierwille (1983) reported significant increases in Cooper-Harper Rating Scale ratings as the levels of wind gust increased and/or as the aircraft pitch stability decreased. Ntuen et al. (1996) reported increases in Cooper-Harper Rating Scale ratings as instability in a compensatory tracking task increased. The highest ratings were for acceleration control; the lowest for position control; rate control was in the middle.

Data requirement – The scale provides ordinal data that must be analyzed accordingly. The Cooper-Harper Rating Scale should be used for workload assessment only if handling difficulty is the major determinant of workload. The task must be fully defined for a common reference.

Thresholds – Ratings vary from 1 (excellent, highly desirable) to 10 (major deficiencies). Noninteger ratings are not allowed.

Sources

Connor, S.A., and Wierwille, W.W. Comparative evaluation of twenty pilot workload assessment measures using a psychomotor task in a moving base aircraft simulator (Report 166457). Moffett Field, CA: NASA Ames Research Center, January 1983.

Cooper, G.E., and Harper, R.P. The use of pilot rating in the evaluation of aircraft handling qualities (AGARD Report 567). London: Technical Editing and Reproduction Ltd., April 1969.

Crabtree, M.S. *Human Factors Evaluation of Several Control System Configurations, Including Workload Sharing with Force Wheel Steering during Approach and Flare (AFFDL-TR-75-43)*. Wright-Patterson Air Force Base, OH: Flight Dynamics Laboratory, April 1975.

Harper, R.P., and Cooper, G.E. Handling qualities and pilot evaluation. AIAA, AHS, ASEE, Aircraft Design Systems and Operations meeting. AIAA Paper 84-2442, 1984.

Krebs, M.J., and Wingert, J.W. *Use of the Oculometer in Pilot Workload Measurement (NASA CR-144951)*. Washington, DC: National Aeronautics and Space Administration, February 1976.

Lebacqz, J.V., and Aiken, E.W. A flight investigation of control, display, and guidance requirements for decelerating descending VTOL instrument transitions using the X-22A variable stability aircraft (AK-5336-F-1). Buffalo, NY: Calspan Corporation, September 1975.

Ntuen, C.A., Park, E., Strickland, D., and Watson, A.R. A frizzy model for workload assessment in complex task situations. IEEE 0-8186-7493: 101–107, 1996.

Schultz, W.C., Newell, F.D., and Whitbeck, R.F. A study of relationships between aircraft system performance and pilot ratings. Proceedings of the 6th Annual NASA University Conference on Manual Control, 339–340, 1970.

Wierwille, W.W., and Connor, S.A. Evaluation of 20 workload measures using a psychomotor task in a moving-base aircraft simulator. *Human Factors* 25(1): 1–16, 1983.

2.3.2.3 Honeywell Cooper-Harper Rating Scale

General description – This rating scale (see Figure 2.7) uses a decision-tree structure for assessing overall task workload.

Strengths and limitations – The Honeywell Cooper-Harper Rating Scale was developed by Wolf (1978) to assess overall task workload. North et al. (1979) used the scale to assess workload associated with various Vertical Take-Off and Landing (VTOL) aircraft displays. For the small subset of conditions analyzed, the scale ratings correlated well with performance.

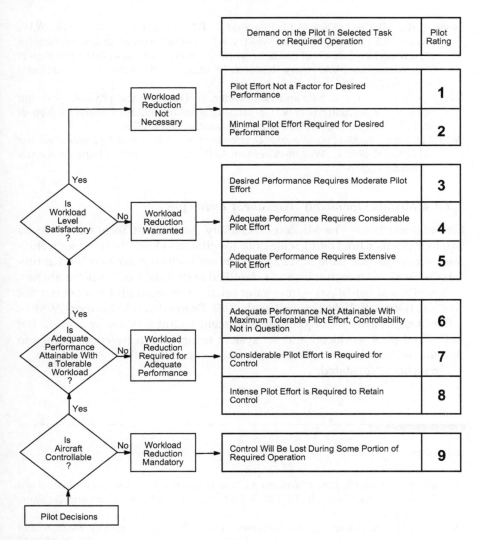

FIGURE 2.7
Honeywell Cooper-Harper Rating Scale (from Lysaght et al., 1989, p.86).

Data requirements – Participants must answer three questions related to task performance. The ratings are ordinal and must be treated as such in subsequent analyses.

Thresholds – Minimum is 1 and the maximum is 9.

Sources

Lysaght, R.J., Hill, S.G., Dick, A.O., Plamondon, B.D., Linton, P.M., Wierwille, W.W., Zaklad, A.L., Bittner, A.C., and Wherry, R.J. Operator workload: Comprehensive review and evaluation of operator workload methodologies (Technical Report 851). Alexandria, VA: Army Research Institute for the Behavioral and Social Sciences, June 1989.

North, R.A., Stackhouse, S.P., and Graffunder, K. Performance, physiological and oculometer evaluations of VTOL landing displays (NASA Contractor Report 3171). Hampton, VA: NASA Langley Research Center, 1979.

Wolf, J.D. *Crew Workload Assessment: Development of a Measure of Operator Workload (AFFDL-TR-78-165)*. Wright-Patterson AFB, OH: Air Force Flight Dynamics Laboratory, 1978.

2.3.2.4 Mission Operability Assessment Technique

General description – The Mission Operability Assessment Technique includes two four-point ordinal rating scales, one for pilot workload, the other for technical effectiveness (see Table 2.18). Participants rate both pilot workload and technical effectiveness for each subsystem identified in the task analysis of the aircraft.

Strengths and limitations – Inter-rater reliabilities are high for most but not all tasks (Donnell, 1979; Donnell et al., 1981; Donnell and O'Connor, 1978).

Data requirements – Conjoint measurement techniques are applied to the individual pilot workload and subsystem technical effectiveness ratings to develop an overall interval scale of systems capability.

Thresholds – Not stated.

Sources

Donnell, M.L. *An application of decision-analytic techniques to the test and evaluation of a major air system Phase III (TR-PR-79-6-91)*. McLean, VA: Decisions and Designs, May 1979.

Donnell, M.L., Adelman, L, and Patterson, J.F. *A Systems Operability Measurement Algorithm (SOMA): Application, Validation, and Extensions (TR-81-11-156)*. McLean, VA: Decisions and Designs, April 1981.

Donnell, M.L., and O'Connor, M.F. *The Application of Decision Analytic Techniques to the Test and Evaluation Phase of the Acquisition of a Major Air System Phase II (TR-78-3-25)*. McLean, VA: Decisions and Designs, April 1978.

O'Donnell, R.D., and Eggemeier, F.T. Workload assessment methodology. In K.R. Boff, L. Kaufman, and J.P. Thomas (Eds.) *Handbook of Perception and Human Performance* (pp. 42-1–42-29). New York: John Wiley, 1986.

2.3.2.5 Modified Cooper-Harper Rating Scale

General description – Wierwille and Casali (1983) noted that the Cooper-Harper Rating Scale represented a combined handling-qualities/workload rating scale. They found that it was sensitive to psychomotor demands on an operator, especially for aircraft handling qualities. They wanted to develop an equally useful scale for the estimation of workload associated with cognitive functions, such as "perception, monitoring, evaluation, communications, and problem solving." The Cooper-Harper Rating Scale terminology was not suited to this purpose. A Modified Cooper-Harper Rating Scale (see Figure 2.8) was developed to "increase the range of applicability to situations commonly found in modern systems." Modifications included: (1) changing the rating-scale end points to very easy and impossible, (2) asking the pilot to rate mental workload level rather than

TABLE 2.18

Mission Operability Assessment Technique Pilot Workload and Subsystem Technical Effectiveness Rating Scales

Pilot Workload
1. The pilot workload (PW)/compensation (C)/interference (I) required to perform the designated task is *extreme*. This is a *poor* rating on the PW/C/I dimension.
2. The pilot workload/compensation/interference required to perform the designated task is *high*. This is a *fair* rating on the PW/C/I dimension.
3. The pilot workload/compensation/interference required to perform the designated task is *moderate*. This is a *good* rating on the PW/C/I dimension.
4. The pilot workload/compensation/interference required to perform the designated task is *low*. This is an *excellent* rating on the PW/C/I dimension.

Subsystem Technical Effectiveness
1. The technical effectiveness of the required subsystem is *inadequate* for performing the designated task. Considerable redesign is necessary to attain task requirements. This is a *poor* rating on the subsystem technical effectiveness scale.
2. The technical effectiveness of the required subsystem is *adequate* for performing the designated task. Some redesign is necessary to attain task requirements. This is a *fair* rating on the subsystem technical effectiveness scale.
3. The technical effectiveness of the required subsystem *enhances individual task performance*. No redesign is necessary to attain task requirements. This is a *good* rating on the subsystem technical effectiveness scale.
4. The technical effectiveness of the required subsystem *allows for the integration of multiple tasks*. No redesign is necessary to attain task requirements. This is an *excellent* rating on the subsystem effectiveness scale (O'Donnell and Eggemeier, 1986, p. 42–16).

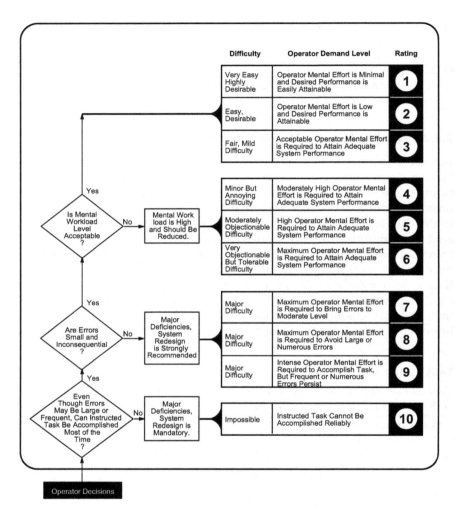

FIGURE 2.8
Modified Cooper-Harper Rating Scale.

controllability, and (3) emphasizing difficulty rather than deficiencies. In addition, Wierwille and Casali (1983) defined mental effort as "minimal" in rating 1, while mental effort is not defined as minimal until rating 3 in the original Cooper-Harper Rating Scale. Further, adequate performance begins at rating 3 in the Modified Cooper-Harper Rating Scale but at rating 5 in the original scale.

Strengths and limitations – Investigations were conducted to assess the Modified Cooper-Harper Rating Scale. They focused on perception (e.g., aircraft engine instruments out of limits during simulated flight), cognition (e.g., arithmetic problem solving during simulated flight), communications (e.g., detection of, comprehension of, and response to own aircraft call sign during simulated flight), and stress. Reliability has been high.

Perception. Itoh et al. (1990) used the Modified Cooper-Harper Rating Scale to compare workload in the Boeing 747-400 electromechanical displays and the Boeing 767 integrated CRT displays. There was no significant difference in workload ratings. In another aircraft application, Jennings et al. (2004) also reported no difference in Modified Cooper Harper Workload Scale ratings of 11 pilots flying a helicopter using a Tactical Situational Awareness System.

Cognition. Wierwille et al. (1985b) reported a significant increase in workload as navigation load increased.

Communication. Casali and Wierwille (1983a, 1983b) reported that Modified Cooper-Harper Rating Scale ratings increased as the communication load increased. Skipper et al. (1986) reported significant increases in ratings in both high communication and high navigation loads. Casto and Casali (2010) reported significantly higher workload ratings as visibility decreased, number of maneuvers increased, and amount of information in a communication increased. The data were from Army helicopter pilots in a Black Hawk simulator.

Stress. Casali and Wierwille (1984) reported significant increases in ratings as the number of danger conditions increased. Wolf (1978) reported the highest workload ratings in the highest workload flight condition (i.e., high wind gust and poor handling qualities).

Reliability. Bittner et al. (1989) reported reliable differences between mission segments in a mobile air defense system. Byers et al. (1988) reported reliable differences between crew positions in a remotely piloted vehicle system. These results suggested that the Modified Cooper-Harper Rating Scale is a valid, statistically reliable indicator of overall mental workload. However, it carries with it the underlying assumptions that high workload is the only determinant of the need for changing the control/display configuration. In spite of that assumption it has been widely used in cockpit evaluations and comparisons.

Wierwille et al. (1985a) concluded that the Modified Cooper-Harper Rating Scale provided consistent and sensitive ratings of workload across a range of tasks. Wierwille et al. (1985c) reported the best consistency and sensitivity with the Modified Cooper-Harper Rating Scale from five alternative tests. Warr et al. (1986) reported that the Modified Cooper-Harper Rating Scale ratings were as sensitive to task difficulty as SWAT ratings. Kilmer et al. (1988), however, reported that the Modified Cooper-Harper Rating Scale was less sensitive than SWAT ratings to changes in tracking task difficulties. Hill et al. (1992) reported that the Modified Cooper-Harper Rating Scale was not as sensitive or as operator accepted as the NASA TLX or the Overall Workload Scale.

Papa and Stoliker (1988) tailored the Modified Cooper-Harper Rating Scale to evaluate the Low Altitude Navigation and Targeting Infrared System for Night (LANTIRN) on an F-16 aircraft.

Data requirements – Wierwille and Casali (1983) recommend the use of the Modified Cooper-Harper Rating Scale in experiments where overall mental workload is to be assessed. They emphasized the importance of proper instructions to the participants. Since the scale was designed for use in experimental situations, it may not be appropriate to situations requiring an absolute diagnosis of a subsystem. Harris et al. (1992) recommend the use of non-parametric analysis techniques since the Modified Cooper-Harper Rating Scale is not an interval scale.

Thresholds – Not stated.

Sources

Bittner, A.C., Byers, J.C., Hill, S.G., Zaklad, A.L., and Christ, R.E. Generic workload ratings of a mobile air defense system (LOS-F-H). Proceedings of the 33rd Annual Meeting of the Human Factors Society, 1476–1480, 1989.

Byers, J.C., Bittner, A.C., Hill, S.G., Zaklad, A.L., and Christ, R.E. Workload assessment of a remotely piloted vehicle (RPV) system. Proceedings of the 32nd Annual Meeting of the Human Factors Society, 1145–1149, 1988.

Casali, J.G., and Wierwille, W.W. A comparison of rating scale, secondary task, physiological, and primary-task workload estimation techniques in a simulated flight task emphasizing communications load. *Human Factors* 25: 623–642, 1983a.

Casali, J.G., and Wierwille, W.W. Communications-imposed pilot workload: A comparison of sixteen estimation techniques. Proceedings of the Second Symposium on Aviation Psychology, 223–235, 1983b.

Casali, J.G., and Wierwille, W.W. On the comparison of pilot perceptual workload: A comparison of assessment techniques addressing sensitivity and intrusion issues. *Ergonomics* 27: 1033–1050, 1984.

Casto, K.L., and Casali, J.G. Effect of communications headset, hearing ability, flight workload, and communications signal quality on pilot performance in an Army Black Hawk helicopter simulator. Proceedings of the Human Factors and Ergonomics Society 54th Annual Meeting, 80–84, 2010.

Harris, R.M., Hill, S.G., Lysaght, R.J., and Christ, R.E. Handbook for operating the OWLKNEST technology (ARI Research Note 92-49). Alexandria, VA: United States Army Research Institute for the Behavioral and Social Sciences, 1992.

Hill, S.G., Iavecchia, H.P., Byers, J.C., Bittner, A.C., Zaklad, A.L., and Christ, R.E. Comparison of four subjective workload rating scales. *Human Factors* 34: 429–439, 1992.

Itoh, Y., Hayashi, Y., Tsukui, I., and Saito, S. The ergonomics evaluation of eye movement and mental workload in aircraft pilots. *Ergonomics* 33(6): 719–733, 1990.

Jennings, S., Craig, G., Cheung, B., Rupert, A., and Schultz, K. Flight-test of a tactile situational awareness system in a land-based deck landing task. Proceedings of the Human Factors and Ergonomics Society 48th Annual Meeting, 142–146, 2004.

Kilmer, K.J., Knapp, R., Burdsal, C., Borresen, R., Bateman, R., and Malzahn, D. Techniques of subjective assessment: A comparison of the SWAT and Modified Cooper-Harper scale. Proceedings of the Human Factors Society 32nd Annual Meeting, 155–159, 1988.

Papa, R.M., and Stoliker, J.R. Pilot workload assessment: A flight test approach (88-2105). Fourth Flight Test Conference. Washington, DC: American Institute of Aeronautics and Astronautics, 1988.

Skipper, J.H., Rieger, C.A., and Wierwille, W.W. Evaluation of decision-tree rating scales for mental workload estimation. *Ergonomics* 29: 585–599, 1986.

Warr, D., Colle, H., and Reid, G. A comparative evaluation of two subjective workload measures: The subjective workload assessment technique and the Modified Cooper-Harper scale. Presented at the Symposium on Psychology in Department of Defense. Colorado Springs, CO: US Air Force Academy, 1986.

Wierwille, W.W., and Casali, J.G. A validated rating scale for global mental workload measurement applications. Proceedings of the 27th Annual Meeting of the Human Factors Society, 129–133, 1983.

Wierwille, W.W., Casali, J.G., Connor, S.A., and Rahimi, M. Evaluation of the sensitivity and intrusion of mental workload estimation techniques. In W. Romer (Ed.) Advances in Man-Machine Systems Research (vol. 2, pp. 51–127). Greenwich, CT: J.A.I. Press, 1985a.

Wierwille, W.W., Rahimi, M., and Casali, J.G. Evaluation of 16 measures of mental workload using a simulated flight task emphasizing mediational activity. *Human Factors* 27(5): 489–502, 1985b.

Wierwille, W.W., Skipper, J., and Reiger, C. *Decision Tree Rating Scales for Workload Estimation: Theme and Variations (N85-11544)*. Blacksburg, VA: Vehicle Simulation Laboratory, 1985c.

Wolf, J.D. *Crew Workload Assessment: Development of a Measure of Operator Workload (AFFDL-TR-78-165)*. Wright-Patterson AFB, OH: Air Force Flight Dynamics Laboratory, December 1978.

2.3.2.6 Sequential Judgment Scale

General description – Pitrella and Kappler (1988) developed the Sequential Judgment Scale to measure the difficulty of driver vehicle handling. It was designed to meet the following rating scale guidelines: "(1) use continuous instead of category scale formats, (2) use both verbal descriptors and numbers at scale points, (3) use descriptors at all major scale markings, (4) use horizontal rather than vertical scale formats, (5) either use extreme or no descriptors at end points, (6) use short, precise, and value-unloaded descriptors, (8) select and use equidistant descriptors, (9) use psychologically-scaled descriptors, (10) use positive numbers only, (11) have desirable qualities increase to the right, (12) use descriptors free of evaluation demands and biases, (13) use 11 or more scale points as available descriptors permit, and (14) minimize rater workload with suitable aids" (Pfendler et al., 1994, p. 28). The scale has interval scale properties. It exists in both 11- and 15-point versions in German, Dutch, and English. The 15-point English version is presented in Figure 2.9.

Strength and limitations – Kappler et al. (1988) reported that the Sequential Judgment Scale ratings varied significantly between loaded and unloaded trucks as well as between different models of trucks. Kappler and Godthelp (1989) reported significantly more difficulty in vehicle handling as tire pressure and lane width decreased. The participants drove on a closed-loop straight lane.

Pitrella (1988) reported that the scale significantly discriminated 10 difficulty levels in a tracking task. Difficulty was manipulated by varying the amplitude and frequency of the forcing function. Pfendler (1993) reported

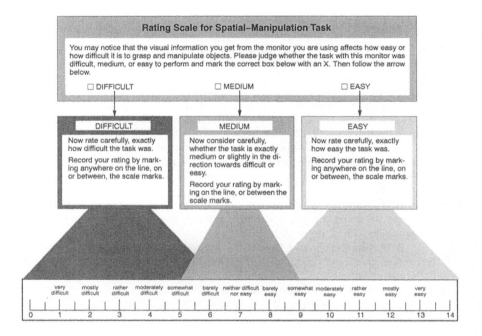

FIGURE 2.9
15-point form of the sequential judgment scale (Pfender et al., 1994, p. 31).

higher validity estimates for the Sequential Judgment Scale (+0.72) than the German-version of the NASA TLX (+0.708) in a color detection task.

Reliabilities have also been high (+0.92 to +0.99, Kappler et al., 1988; +0.87 to +0.99, Pitrella, 1988; +0.87, Pfendler, 1993). Since the scale is an interval scale, parametric statistics can be used to analyze the data.

The scale has two disadvantages: "(1) if only overall workload is measured, rating results will have low diagnosticity and (2) information on validity of the scale is restricted to psychomotor and perceptual tasks" (Pfendler et al., 1994, p. 30).

Data requirements – Participants mark the scale in pen or pencil. The experimenter then measures the distance from the right end of the scale. This measure is converted to a percentage of the complete scale.

Thresholds – 0% to 100%.

Sources

Kappler, W.D., and Godthelp, H. Design and use of the two-level Sequential Judgment Rating Scale in the identification of vehicle handling criteria: I. Instrumented car experiments on straight lane driving (FAT Report Number 79). Wachtberg: Forschungsinstitut fur Anthropotechnik, 1989.

Kappler, W.D., Pitrella, F.D., and Godthelp, H. Psychometric and performance measurement of light weight truck handling qualities (FAT Report Number 77). Wachtberg: Forschungsinstitut fur Anthropotechnik, 1988.

Pfendler, C. Vergleich der Zwei-Ebenen Intensitats-Skala und des NASA Task Load Index bei de Beanspruchungsbewertung wahrend ternvorgangen. Z. Arb. Wise 47 (19 NF) 1993/1, 26–33.

Pfendler, C., Pitrella, F.D., and Wiegand, D. Workload measurement in human engineering test and evaluation. Forschungsinstitut fur Anthropotechnik. Bericht Number 109, July 1994.

Pitrella, F.D. A cognitive model of the internal rating process (FAT Report Number 82). Wachtberg: Forschungsinstitut fur Anthropotechnik, 1988.

Pitrella, F.D., and Kappler, W.D. Identification and evaluation of scale design principles in the development of the sequential judgment, extended range scale (FAT Report Number 80). Wachtberg: Forschungsinstitut fur Anthropotechnik, 1988.

2.3.3 Set of Subscales Subjective Workload Measures

The final type of subjective workload measure is a set of subscales each of which was designed to measure different aspects of workload. Examples include: Assessing the Impact of Automation on Mental Workload (AIM-s) (Section 2.3.3.1), Crew Status Survey which separates fatigue and workload (Section 2.3.3.2), Finegold Workload Rating Scale (Section 2.3.3.3), Flight Workload Questionnaire (Section 2.3.3.4), Hart and Hauser Rating Scale (Section 2.3.3.5), Human Robot Interaction Workload Measurement Tool (Section 2.3.3.6), Multi-Descriptor Scale (Section 2.3.3.7), Multidimensional Rating Scale (Section 2.3.3.8), Multiple Resources Questionnaire (Section 2.3.3.9), NASA Bipolar Rating Scale (Section 2.3.3.10), NASA Task Load Index (Section 2.3.3.11), Profile of Mood States (Section 2.3.3.12), Subjective Workload Assessment Technique (Section 2.3.3.13), Team Workload Questionnaire (Section 2.3.3.14), Workload/Compensation/Interference/Technical Effectiveness (Section 2.3.3.15).

2.3.3.1 Assessing the Impact of Automation on Mental Workload (AIM)

General description – The Assessing the Impact of Automation on Mental Workload (AIM-s) questionnaire was designed to measure the workload of Air Traffic Controllers. There are two versions: AIM-s (short) which contains one question – "In the previous working period(s), how much effort did it take to …". This is followed by a list of tasks, each of which has a rating scale of 0 to 6 (see Figure 2.10). The longer version is AIM-I. It has 32 questions, four in each of the following eight areas: (1) building and maintaining situational awareness, (2) monitoring of information sources, (3) memory management, (4) managing the controller working position, (5) diagnosing and problem detection, (6) decision making and problem solving, (7) resource management and multitasking, and (8) team awareness.

	none	very little	little	some	much	very much	extreme
1) ... prioritise tasks?	0	1	2	3	4	5	6
2) ... identify potential conflicts?	0	1	2	3	4	5	6
3) ... scan radar or any display?	0	1	2	3	4	5	6
4) ... evaluate conflict resolution options against the traffic situation and any constraints?	0	1	2	3	4	5	6
5) ... anticipate the future traffic situation?	0	1	2	3	4	5	6
6) ... recognise a mismatch of available data with my traffic picture?	0	1	2	3	4	5	6
7) ... issue commands in time?	0	1	2	3	4	5	6
8) ... evaluate the consequences of a plan?	0	1	2	3	4	5	6
9) ... manage flight data information?	0	1	2	3	4	5	6
10) ... share information with team members?	0	1	2	3	4	5	6
11) ... recall necessary information?	0	1	2	3	4	5	6
12) ... anticipate team members' needs?	0	1	2	3	4	5	6
13) ... prioritise requests?	0	1	2	3	4	5	6
14) ... scan flight progress data?	0	1	2	3	4	5	6
15) ... access relevant aircraft or flight information?	0	1	2	3	4	5	6
16) ... gather and interpret information?	0	1	2	3	4	5	6

FIGURE 2.10
AIM-s (http://www.eurocontrol.int/humanfactors/public/standard_page/SHAPE_Questionnaires.html).

Strengths and limitations – Dehn (2008) described the steps taken in the development of the AIM: (1) literature review to obtain initial set of items, (2) requirement-based review for easy administration, ease of understanding, consistent format, and scoring key provided, (3) collection of expert feedback, and (4) initial empirical study. The study participants were 24 active Air Traffic Controllers. Items were eliminated from the questionnaire if they reduced internal consistency or were redundant.

The European Organization for the Safety for Air Navigation recommended using AIM-s for screening purposes to compare two or more systems. They further recommended that the AIM-s be completed at least twice daily. Finally, they warn that there does not yet exist an extensive set of norm data nor has a formal validation been completed.

Data requirements – For AIM-s, participants indicate workload using the rating scale shown above. For AIM-I, participants answer 32 questions. Scoring sheet for both versions of AIM are available from SHAPE.

Thresholds – 0% to 100%.

Source

Dehn, D.M. Assessing the impact of automation on the Air Traffic Controller: The SHAPE questionnaires. *Air Traffic Control Quarterly* 16(2): 127–146, 2008.

2.3.3.2 Crew Status Survey

General description – The original Crew Status Survey was developed by Pearson and Byars (1956) and contained 20 statements describing fatigue states. The staff of the Air Force School of Aerospace Medicine Crew Performance Branch, principally Storm and Parke, updated the original survey. They selected the statements anchoring the points on the fatigue scale of the survey through iterative presentations of drafts of the survey to aircrew members. The structure of the fatigue scale was somewhat cumbersome, since the dimensions of workload, temporal demand, system demand, system management, danger, and acceptability were combined on one scale. However, the fatigue scale was simple enough to be well received by operational crews. The fatigue scale of the survey was shortened to seven statements and subsequently tested for sensitivity to fatigue as well as for test/retest reliability (Miller and Narvaez, 1986). Finally, a seven-point workload scale was added. The current Crew Status Survey (see Figure 2.11) provides measures of self-reported fatigue and workload as well as space for general comments. Ames and George (1993) modified the workload scale to enhance reliability. Their scale descriptors are:

1. Nothing To Do; No System Demands.
2. Light Activity; Minimum Demands.
3. Moderate Activity; Easily Managed; Considerable Spare Time.
4. Busy; Challenging But Manageable; Adequate Time Available.
5. Very Busy; Demanding To Manage; Barely Enough Time.
6. Extremely Busy; Very Difficult; Non-Essential Tasks Postponed.
7. Overloaded; System Unmanageable; Important Tasks Undone; Unsafe. (p. 4).

NAME	DATE AND TIME

SUBJECT FATIGUE

(Circle the number of the statement which describes how you feel RIGHT NOW.)

1	Fully Alert, Wide Awake; Extremely Peppy
2	Very Lively; Responsive, But Not at Peak
3	Okay; Somewhat Fresh
4	A Little Tired; Less Than Fresh
5	Moderately Tired; Let Down
6	Extremely Tired; Very Difficult to Concentrate
7	Completely Exhausted; Unable to Function Effectively; Ready to Drop

COMMENTS

WORKLOAD ESTIMATE

(Circle the number of the statement which describes the MAXIMUM workload you experienced during the past work period. Put an X over the number of the statement which best describes the AVERAGE workload you experienced during the past work period.)

1	Nothing to do; No System Demands
2	Little to do; Minimum System Demands
3	Active Involvement Required, But Easy to Keep Up
4	Challenging, But Manageable
5	Extremely Busy; Barely Able to Keep Up
6	Too Much to do; Overloaded; Postponing Some Tasks
7	Unmanageable; Potentially Dangerous; Unacceptable

COMMENTS

SAM FORM 202   CREW STATUS SURVEY

FIGURE 2.11
Crew status survey.

Strengths and limitations – These scales have been found to be sensitive to changes in task demand and fatigue as well as display type (Ellis et al., 2011) but are independent of each other (Courtright et al., 1986). Storm and Parke (1987) used the Crew Status Survey to assess the effects of temazepam on FB-111A crewmembers. The effect of the drug was not significant. The effect of performing the mission was, however. Specifically, the fatigue ratings were higher at the end than at the beginning of a mission. Gawron et al. (1988) analyzed Crew Status Survey ratings made at four times during each flight. They found a significant segment effect on fatigue and workload. Fatigue ratings increased over the course of the flight (preflight = 1.14, predrop = 1.47, postdrop = 1.43, and postflight = 1.56). Workload ratings were highest around a simulated air drop (preflight = 1.05, predrop = 2.86, postdrop = 2.52, and postflight = 1.11).

George et al. (1991) collected workload ratings from Combat Talon II aircrew members during arctic deployment. None of the median ratings were greater than four. However, level 5 ratings occurred for navigators during airdrops and self-contained approach run-ins. These authors also used the Crew Status Survey workload scale during terrain-following training flights on Combat Talon II. Pilots and copilots gave a median rating of 7. The ratings were used to identify major crewstation deficiencies.

However, George and Hollis (1991) reported confusion between adjacent categories at the high workload end of the Crew Status Survey. They also found adequate ordinal properties for the scale but very large variance in most order-of-merit tables. In a more recent study, Thomas (2011) used the Revised Crew Status Survey to evaluate automation in a commercial aircraft flightdeck (see Table 2.19). She reported that increasing automation level significantly decreased pilot workload ratings. She points out, however, that the practical difference is small (2 versus 3).

Data requirements – Although the Crew Status Survey is printed on card stock, participants find it difficult to fill in the rating scale during high workload periods. Further, sorting (for example, by the time completed) the completed card-stock ratings after the flight is also difficult and not error free. A larger character-size version of the survey has been included on flight cards at the Air Force Flight Test Center. Verbal ratings prompted by the experimenter

TABLE 2.19

Revised Crew Status Survey (Thomas, 2011, p. 13)

Rating	Definition
1	Nothing to do; No system demands
2	Light activity; Minimum demands
3	Moderate activity; Easily managed; Considerable spare time
4	Busy; Challenging but manageable; Adequate time available
5	Very busy; Demanding to manage; Barely enough time
6	Extremely busy; Very difficult; Non-essential tasks postponed
7	Overloaded; System unmanageable; Essential tasks undone; Unsafe

work well if: (1) participants can quickly scan a card-stock copy of the rating scale to verify the meaning of a rating and (2) participants are not performing a conflicting verbal task. Each scale can be used independently.

Thresholds – 1 to 7 for subjective fatigue; 1 to 7 for workload.

Sources

Ames, L.L., and George, E.J. *Revision and Verification of a Seven-Point Workload Estimate Scale (AFFTC-TIM-93-01)*. Edwards Air Force Base, CA: Air Force Flight Test Center, 1993.

Courtright, J.F., Frankenfeld, C.A., and Rokicki, S.M. The independence of ratings of workload and fatigue. Paper presented at the Human Factors Society 30th Annual Meeting, 1986.

Ellis, K.K.E., Kramer, L.J., Shelton, K.J., Arthur, J.J., and Prinzel, L.J. Transition of attention in terminal area NextGen operations using synthetic vision systems. Proceedings of the Human Factors and Ergonomics Society 55th Annual Meeting, 46–50, 2011.

Gawron, V.J., Schiflett, S.G., Miller, J., Ball, J., Slater, T., Parker, F., Lloyd, M., Travale, D., and Spicuzza, R.J. *The Effect of Pyridostigmine Bromide on In-Flight Aircrew Performance (USAFSAM-TR-87-24)*. Brooks Air Force Base, TX: School of Aerospace Medicine, January 1988.

George, E., and Hollis, S. *Scale Validation in Flight Test*. Edwards Air Force Base, CA: Flight Test Center, December 1991.

George, E.J., Nordeen, M., and Thurmond, D. *Combat Talon II Human Factors Assessment (AFFTC TR 90-36)*. Edwards Air Force Base, CA: Flight Test Center, November 1991.

Miller, J.C., and Narvaez, A. A comparison of two subjective fatigue checklists. Proceedings of the 10th Psychology in the DoD Symposium, 514–518, 1986.

Pearson, R.G., and Byars, G.E. *The Development and Validation of a Checklist for Measuring Subjective Fatigue (TR-56-115)*. Brooks Air Force Base, TX: School of Aerospace Medicine, 1956.

Storm, W.F., and Parke, R.C. FB-111A aircrew use of temazepam during surge operations. Proceedings of the NATO Advisory Group for Aerospace Research and Development (AGARD) Biochemical Enhancement of Performance Conference (Paper number 415). Neuilly-sur-Seine, France: AGARD, 12-1–12-12, 1987.

Thomas, L.C. Pilot workload under non-normal event resolution: Assessment of levels of automation and a voice interface. Proceedings of the Human Factors and Ergonomics Society 55th Annual Meeting, 11–15, 2011.

2.3.3.3 Finegold Workload Rating Scale

General description – The Finegold Workload Rating Scale has five subscales (see Figure 2.12). It was developed to evaluate workload at each crewstation aboard the AC-130H Gunship.

Strengths and limitations – Finegold et al. (1986) reported lower ratings associated with cruise than with engagement or threat segments. Analysis of the

Human Workload

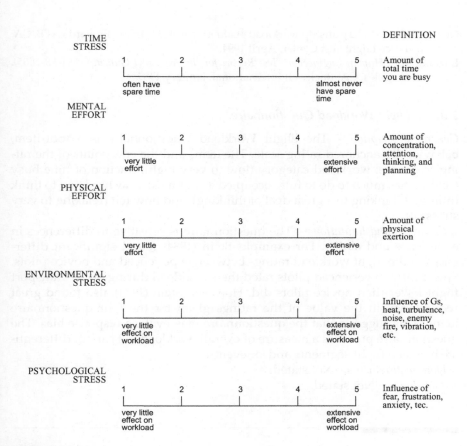

FIGURE 2.12
Finegold Workload Rating Scale.

subscales indicated that time stress was rated differently at each crewstation. Lozano (1989) replicated the Finegold et al. (1986) test using the AC-130 H Gunship. Again, ratings on subscales varied by crew position. George (1994) replicated both studies using the current version of the AC-130U Gunship.

Data requirements – Average individual subscales as well as the complete workload Rating Scale scores.

Thresholds – 1 for low workload and 5 for high workload.

Sources

Finegold, L.S., Lawless, M.T., Simons, J.L., Dunleavy, A.O., and Johnson, J. Estimating crew performance in advanced systems. Volume II: Application to future gunships. Appendix B: Results of data analysis for AC-13OH and hypothetical AC-130H (RP). Edwards Air Force Base, CA: Air Force Flight Test Center, October 1986.

George, E.J. *AC-130U gunship workload evaluation (C4654-3-501)*. Edwards AFB, CA: Air Force Flight Test Center, April 1994.

Lozano, M.L. *Human Engineering Test Report for the AC-130U Gunship (NA-88-1805)*. Los Angeles, CA: Rockwell International, January 1989.

2.3.3.4 Flight Workload Questionnaire

General description – The Flight Workload Questionnaire is a four-item, behaviorally-anchored rating scale. The items and the end points of the rating scales are: workload category (low to very high), fraction of time busy (seldom has much to do to fully occupied at all times), how hard had to think (minimal thinking to a great deal of thinking), and how felt (relaxing to very stressful).

Strengths and limitations – The questionnaire is sensitive to differences in experience and ability. For example, Stein (1984) found significant differences in the flight workload ratings between experienced and novice pilots. Specifically, experienced pilots rated their workload during an air transport flight lower than novice pilots did. However, Stein (1984) also found great redundancy in the value of the ratings given for the four questionnaire items. This suggests that the questionnaire may evoke a response bias. The questionnaire provides a measure of overall workload but cannot differentiate between flight segments and/or events.

Data requirements – Not stated.

Thresholds – Not stated.

Source

Stein, E.S. *The Measurement of Pilot Performance: A Master-Journeyman Approach (DOT/FAA/CT-83/15)*. Atlantic City, NJ: Federal Aviation Administration Technical Center, May 1984.

2.3.3.5 Hart and Hauser Rating Scale

General description – Hart and Hauser (1987) used a six-item rating scale (see Figure 2.13) to measure workload during a nine-hour flight. Participants were instructed to mark the scale position that represented their experience.

Strengths and limitations – The scale was developed for use in flight. In the initial study, Hart and Hauser (1987) asked participants to complete the questionnaire at the end of each of seven flight segments. They reported significant segment effects in the seven-hour flight. Specifically, stress, mental/sensory effort, and time pressure were lowest during a data-recording segment. There was a sharp increase in rated fatigue after the start of the data-recording segment. Overall, the aircraft commander rated workload as

Human Workload

Stress

Completely Relaxed _____ Extremely Tense

Mental/Sensory Effort

Very Low _____ Very High

Fatigue

Wide Awake _____ Worn Out

Time Pressure

None _____ Very Rushed

Overall Workload

Very Low _____ Very High

Performance

Completely Unsatisfactory _____ Completely Satisfactory

FIGURE 2.13
Hart and Hauser Rating Scale.

higher than the copilot did. Finally, performance received the same ratings throughout the flight.

Data requirements – The scale is simple to use but requires a stiff writing surface and minimal turbulence.

Thresholds – Not stated.

Source

Hart, S.G., and Hauser, J.R. Inflight application of three pilot workload measurement techniques. *Aviation, Space, and Environmental Medicine* 58: 402–410, 1987.

2.3.3.6 Human Robot Interaction Workload Measurement Tool

General description – The Human Robot Interaction Workload Measurement Tool (HRI-WM) (Yagoda, 2010) requires participants to rate from lowest to highest the amount of workload in configuration, task, context, team process, and system as well as overall workload from minimum (1) to maximum (5). Participants are also able to provide comments on their rating.

Strengths and limitations – Yagoda (2010) reported a two-phase development process in which the interface for the computerized version of the HRI-WM was enhanced based on participant feedback and usability testing. The testing consisted of 10 graduate and undergraduate students using the HRI-WM to rate the workload of a search and rescue task.

Data requirements – The six rating scales must be presented to the participant with the ability to accept comments.

Source

Yagoda, R.E. Development of the Human Robot Interaction Workload Measurement Tool (HRI-WM). Proceedings of the Human Factors and Ergonomics Society 54th Annual Meeting, 304–308, 2010.

2.3.3.7 Multi-Descriptor Scale

General description – The Multi-descriptor (MD) scale is composed of six descriptors: (1) attentional demand, (2) error level, (3) difficulty, (4) task complexity, (5) mental workload, and (6) stress level. Each descriptor is rated after a task is completed. The MD score is the average of the six descriptor ratings.

Strengths and limitations – Wierwille et al. (1985) reported the MD scores were not sensitive to variations in difficulty of mathematical calculations performed during a simulated flight task.

Data requirements – The six rating scales must be presented to the participant after a flight and the average of the resultant ratings calculated.

Source

Wierwille, W.W., Rahimi, M., and Casali, J.G. Evaluation of 16 measures of mental workload using a simulated flight task emphasizing mediational activity. *Human Factors* 27(5): 489–502, 1985.

2.3.3.8 Multidimensional Rating Scale

General description – The Multidimensional Rating Scale is composed of eight bipolar scales (see Table 2.20). Participants are asked to draw a horizontal line on the scale to indicate their rating.

Strengths and limitations – Damos (1985) reported high correlations among several of the subscales (+0.82) as well as between overall workload and task

Human Workload

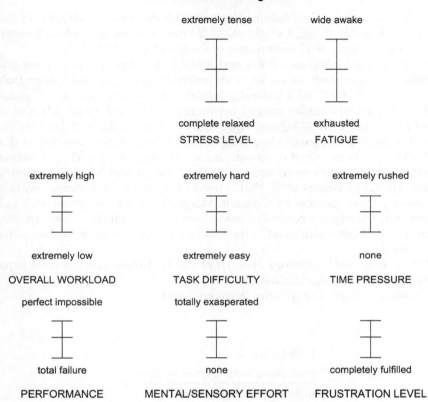

TABLE 2.20
Multidimensional Rating Scale

difficulty single-task condition (+0.73). The time pressure and overall workload scales were significantly associated with task-by-pacing condition interactions. The mental/sensory effort scale was significantly associated with a task by behavior pattern interaction.

Data requirements – The vertical line in each scale must be 100 mm long. The rater must measure the distance from the bottom of the scale to the participant's horizontal line to determine the rating.

Thresholds – Zero to 100.

Source

Damos, D. The relation between the Type A behavior pattern, pacing, and subjective workload under single- and dual-task conditions. *Human Factors* 27(6): 675–680, 1985.

2.3.3.9 Multiple Resources Questionnaire

General description – The Multiple Resources Questionnaire (MRQ) uses one rating scale (0 = no usage, 1 = light usage, 2 = moderate usage, 3 = heavy usage, 4 = extreme usage) for 17 dimensions of workload (see Table 2.21).

Strengths and limitations – Boles and Adair (2001) reported inter-rater reliabilities of undergraduate students for seven computer games to range from $r = +0.67$ to $r = +0.83$. In a second study using just two computer games, the inter-rater reliabilities ranged between $r = +0.57$ and $+0.65$. Klein et al. (2009) used both the MRQ and the NASA TLX to evaluate 2-dimensional and 3-dimensional viewing in a surgical robot. The authors concluded that "The MRQ data provided diagnostic information regarding which information processing pools were stressed in both the 2d and 3d viewing conditions" (p. 1186). Sellers et al. (2012) used MRQ to evaluate three levels of autonomy (Management by Consent, Management by Exception, and Full Autonomy) during the control of unmanned ground vehicles. They reported significantly higher workload in the first two levels of autonomy than in the third level.

Data requirements – Ratings on each of the 17 dimensions are used separately and without transformation.

Thresholds – Each rating varies between 0 and 4.

TABLE 2.21

Multiple Resources Questionnaire

Auditory emotional process
Auditory linguistic process
Facial figural process
Facial motive process
Manual process
Short-term memory process
Spatial attentive process
Spatial categorical process
Spatial concentrative process
Spatial emergent process
Spatial positional process
Spatial quantitative process
Tactile figural process
Visual lexical process
Visual phonetic process
Visual temporal process
Vocal process

Sources

Boles, D.B., and Adair, L.P. The Multiple Resources Questionnaire (MRQ). Proceedings of the Human Factors and Ergonomics Society 45th Annual Meeting, 1790–1794, 2001.

Klein, M.I., Lio, C.H., Grant, R., Carswell, C.M., and Strup, S. A mental workload study on the 2d and 3d viewing conditions of the da Vinci Surgical Robot. Proceedings of the Human Factors and Ergonomics Society 53rd Annual Meeting, 1186–1190, 2009.

Sellers, B.C., Fincannon, T., and Jentsch, F. The effects of autonomy and cognitive abilities on workload and supervisory control of unmanned systems. Proceedings of the Human Factors and Ergonomics Society 56th Annual Meeting, 1039–1043, 2012.

2.3.3.10 NASA Bipolar Rating Scale

General description – The NASA Bipolar Rating Scale has 10 subscales. The titles, endpoints, and descriptions of each scale are presented in Table 2.22 and the scale itself, in Figure 2.14. If a scale is not relevant to a task, it is given a weight of zero (Hart et al., 1984). A weighting procedure is used to enhance intrasubject reliability by 50% (Miller and Hart, 1984).

Strengths and limitations – The scale has been used in aircraft simulators and laboratories.

Aircraft Simulators. The scale is sensitive to flight difficulty. Bortolussi et al. (1986) reported significant differences in the bipolar ratings between an easy and a difficult flight scenario. Bortolussi et al. (1987) and Kantowitz et al. (1984) reported similar results. Bortolussi et al. (1986) and Bortolussi et al. (1987) reported that the bipolar scales discriminated two levels of difficulty in a motion-based simulator task. Vidulich and Bortolussi (1988) reported significant increases in the overall workload rating from cruise to combat phase in a simulated helicopter. There was no effect of control configuration.

However, Haworth et al. (1986) reported that, although the scale discriminated control configurations in a single-pilot configuration, it did not do so in a pilot/copilot configuration. They also reported correlations of +0.79 with Cooper-Harper Rating Scale ratings and +0.67 with SWAT ratings in a helicopter nap-of-the-earth mission.

Laboratories. Biferno (1985) reported a correlation between workload and fatigue ratings for a laboratory study. Vidulich and Pandit (1986) reported that the bipolar scales discriminated levels of training in a category search task. Vidulich and Tsang (1985a, 1985b, 1985c, 1986) reported that the NASA Bipolar Rating Scale was sensitive to task demand, had

TABLE 2.22

NASA Bipolar Rating Scale Descriptions

Title	Endpoints	Descriptions
Overall Workload	Low, High	The total workload associated with the task considering all sources and components.
Task Difficulty	Low, High	Whether the task was easy, demanding, simple or complex, exacting or forgiving.
Time Pressure	None, Rushed	The amount of pressure you felt due to the rate at which the task elements occurred. Was the task slow and leisurely or rapid and frantic.
Performance	Perfect, Failure	How successful you think you were in doing what we asked you to do and how satisfied you were with what you accomplished.
Mental/Sensory Effort	None, Impossible	The amount of mental and/or perceptual activity that was required (e.g., thinking, deciding, calculating, remembering, looking, searching, etc.).
Physical Effort	None, Impossible	The amount of physical activity that was required (e.g., pushing, pulling, turning, controlling, activating, etc.).
Frustration Level	Fulfilled, Exasperated	How insecure, discouraged, irritated, and annoyed versus secure, gratified, content and complacent you felt.
Stress Level	Relaxed, Tense	How anxious, worried, uptight, and harassed or calm, tranquil, placid, and relaxed you felt.
Fatigue	Exhausted, Alert	How tired, weary, worn out, and exhausted or fresh, vigorous, and energetic you felt.
Activity Type	Skill based, Rule based, Knowledge based	The degree to which the task required mindless reaction to well-learned routines or required the application of known rules or required problem solving and decision making.

Source: From Lysaght et al. (1989) p. 91.

higher inter-rater reliability than SWAT, and required less time to complete than SWAT.

Data requirements – The number of times a dimension is selected by a participant is used to weight each scale. These weights are then multiplied by the scale score, summed, and divided by the total weight to obtain a workload score. The minimum workload value is zero; the maximum, 100. The scale provides a measure of overall workload but is not sensitive to short-term demands. Further, the activity-type dimension must be carefully explained to pilots before use in flight.

Thresholds – Not stated.

Human Workload

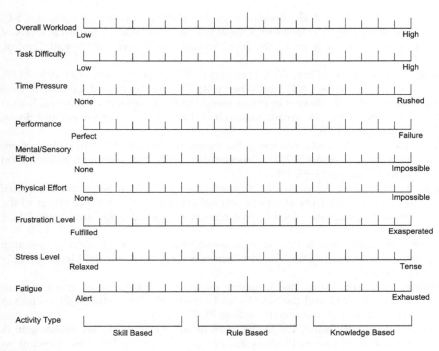

FIGURE 2.14
NASA Bipolar Rating Scale.

Sources

Biferno, M.H. *Mental Workload Measurement: Event-Related Potentials and Ratings of Workload and Fatigue (NASA CR-177354)*. Washington, DC: NASA, 1985.

Bortolussi, M.R., Hart, S.G., and Shively, R.J. Measuring moment-to-moment pilot workload using synchronous presentations of secondary tasks in a motion-base trainer. Proceedings of the 4th Symposium on Aviation Psychology, 651–657, 1987.

Bortolussi, M.R., Kantowitz, B.H., and Hart, S.G. Measuring pilot workload in a motion base trainer: A comparison of four techniques. *Applied Ergonomics* 17: 278–283, 1986.

Hart, S.G., Battiste, V., and Lester, P.T. POPCORN: A supervisory control simulation for workload and performance research (NASA-CP-2341). Proceedings of the 20th Annual Conference on Manual Control, 431–453, 1984.

Haworth, L.A., Bivens, C.C., and Shively, R.J. An investigation of single-piloted advanced cockpit and control configuration for nap-of-the-earth helicopter combat mission tasks. Proceedings of the 42nd Annual Forum of the American Helicopter Society, 675–671, 1986.

Kantowitz, B.H., Hart, S.G., Bortolussi, M.R., Shively, R.J., and Kantowitz, S.C. Measuring pilot workload in a moving-base simulator: II. Building levels of workload. Proceedings of the 20th Annual Conference on Manual Control, 373–396, 1984.

Lysaght, R.J., Hill, S.G., Dick, A.O., Plamondon, B.D., Linton, P.M., Wierwille, W.W., Zaklad, A.L., Bittner, A.C., and Wherry, R.J. Operator workload: Comprehensive review and evaluation of operator workload methodologies (Technical Report 851). Alexandria, VA: Army Research Institute for the Behavioral and Social Sciences, June 1989.

Miller, R.C., and Hart, S.G. Assessing the subjective workload of directional orientation tasks (NASA-CP-2341). Proceedings of the 20th Annual Conference on Manual Control, 85–95, 1984.

Vidulich, M.A., and Bortolussi, M.R. Speech recognition in advanced rotorcraft: Using speech controls to reduce manual control overload. Proceedings of the National Specialists' Meeting Automation Applications for Rotorcraft, 20–29, 1988.

Vidulich, M.A., and Pandit, P. Training and subjective workload in a category search task. Proceedings of the Human Factors Society 30th Annual Meeting, 1133–1136, 1986.

Vidulich, M.A., and Tsang, P.S. Assessing subjective workload assessment: A comparison of SWAT and the NASA-Bipolar methods. Proceedings of the Human Factors Society 29th Annual Meeting, 71–75, 1985a.

Vidulich, M.A., and Tsang, P.S. Techniques of subjective workload assessment: A comparison of two methodologies. Proceedings of the Third Symposium on Aviation Psychology, 239–246, 1985b.

Vidulich, M.A., and Tsang, P.S. Evaluation of two cognitive abilities tests in a dual-task environment. Proceedings of the 21st Annual Conference on Manual Control, 12.1–12.10, 1985c.

Vidulich, M.A., and Tsang, P.S. Techniques of subjective workload assessment: A comparison of SWAT and NASA-Bipolar methods. *Ergonomics* 29(11): 1385–1398, 1986.

2.3.3.11 NASA Task Load Index

General description – The NASA Task Load Index (TLX) is a multi-dimensional subjective workload rating technique (see Figure 2.15). In NASA TLX, workload is defined as the "cost incurred by human operators to achieve a specific level of performance." The subjective experience of workload is defined as an integration of the weighted subjective responses (emotional, cognitive, and physical) and weighted evaluation of behaviors. The behaviors and subjective responses, in turn, are driven by perceptions of task demand. Task demands can be objectively quantified in terms of magnitude and importance. An experimentally based process of elimination led to the identification of six dimensions for the subjective experience of workload: mental demand, physical demand, temporal demand, perceived

Human Workload

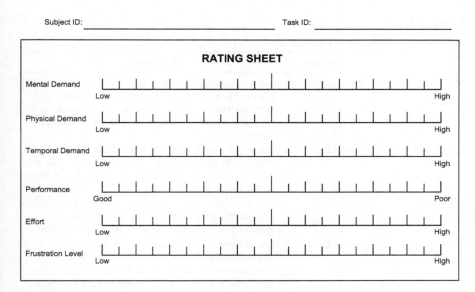

FIGURE 2.15
NASA TLX Rating Sheet.

performance, effort, and frustration level. The rating-scale definitions are presented in Table 2.23.

Strengths and limitations – During the development of the NASA TLX, 16 investigations were carried out, establishing a database of 3,461 entries from 247 participants (Hart and Staveland, 1987). All dimensions were rated on bipolar scales ranging from 1 to 100, anchored at each end with a single adjective. An overall workload rating was determined from a weighted combination of scores on the six dimensions. The weights were determined from a set of relevance ratings provided by the participants.

Hart and Staveland (1987) concluded that the NASA TLX provides a sensitive indicator of overall workload as it differed among tasks of various cognitive and physical demands. They also stated that the weights and magnitudes determined for each NASA TLX dimension provide important diagnostic information about the sources of loading within a task. They reported that the six NASA TLX ratings took less than a minute to acquire and suggested the scale would be useful in operational environments.

Since its development, the NASA TLX has been used to evaluate workload in aircraft, Air Traffic Control, automobiles, unmanned systems, nuclear power plants, laboratories, and healthcare. There have also been a series of reliability studies as well as comparisons to other workload measures.

<u>Aircraft</u>. NASA TLX has been used extensively to evaluate workload in aircraft (Bittner et al., 1989; Byers et al., 1988); Hill et al., 1988, 1989; Lee and

TABLE 2.23
NASA TLX Rating-Scale Descriptions

Title	Endpoints	Descriptions
Mental Demand	Low, High	How much mental and perceptual activity was required (e.g., thinking, deciding, calculating, remembering, looking, searching, etc.)? Was the task easy or demanding, simple or complex, exacting or forgiving?
Physical Demand	Low, High	How much physical activity was required (e.g., pushing, pulling, turning, controlling, activating, etc.)? Was the task easy or demanding, slow or brisk, slack or strenuous, restful or laborious?
Temporal Demand	Low, High	How much time pressure did you feel due to the rate or pace at which the tasks or task elements occurred? Was the pace slow and leisurely or rapid and frantic?
Performance	Good, Poor	How successful do you think you were in accomplishing the goals of the task set by the experimenter (or yourself)? How satisfied were you with your performance in accomplishing these goals?
Effort	Low, High	How hard did you have to work (mentally and physically) to accomplish your level of performance?
Frustration Level	Low, High	How insecure, discouraged, irritated, stressed, and annoyed versus secure, gratified, content, relaxed and complacent did you feel during the task? (NASA Task Load Index, p. 13)

Liu (2003); and Fern et al., 2011). Workload comparisons have been made across controls, displays, flight segments, and systems.

Displays In a display study, Stark et al. (2001) reported a significant decrease in workload when pilots in a fixed-base simulator had a tunnel or pathway in the sky display or a smaller rather than a larger display. Alexander et al. (2009), however, reported that NASA TLX indicated highest workload in a medium cluttered Head Up Display (HUD) than in the low or high cluttered HUD. Further, there were no significant differences in NASA TLX ratings due to phase of approach to an airport or in wind.

Grubb et al. (1995), based on vigilance data from 144 observers, reported increased workload as display uncertainty increased. There was no significant effect of period of watch (10, 20, 30, or 40 minutes). Rodes and Gugerty (2012) reported significantly higher workload scores on the NASA TLX for a north up rather than for a track up navigation display.

Flight segments In an early in-flight evaluation in the NASA Kuiper Airborne Observatory, Hart et al. (1984) reported a significant difference between left and right seat positions as well as in flight segments related to ratings of overall fatigue, time pressure, stress, mental/sensory effort, fatigue, and performance. The data were from nine NASA test pilots over 11 flights. Hart and Staveland (1987), based on in-flight data, stated that NASA TLX ratings significantly discriminated flight segments. Nygren (1991) reported that NASA TLX is a measure of general workload experienced by aircrews.

Hancock et al. (1995) reported the NASA TLX score was highly correlated with difficulty of a simulated flight task.

Vidulich and Bortolussi (1988a) reported a significant flight-segment effect but reported no significant differences in NASA TLX ratings between control configurations or between combat countermeasure conditions. Vidulich and Bortolussi (1988b) reported significant increases in NASA TLX ratings from the cruise to the combat phase during a simulated helicopter mission. There was no effect of control configuration, however.

Kennedy et al. (2014) reported that participants who detected a runway incursion during a simulated landing did not have significantly different scores from the participants who did not detect the incursion. The participants were 60 nonpilots aged 20 to 64.

Systems Casner (2009) also reported a significant phase of flight effects for all seven measures of workload in the NASA TLX for the advanced cockpit configuration and in all but one measure for the conventional cockpit. There was also a significant interaction effect for VOR versus GPS, specifically, GPS was associated with higher workload during setup but lower workload during the missed approach phase. There were no significant differences, however, between manual and autopilot control or between conventional versus electronic instruments. Brown and Galster (2004) reported no significant effects of imposed workload on NASA TLX rating. The imposed workload was differences in reliability of automation in a simulated flight.

In a flight profile experiment in a A330 flight simulator, Elmenhorst et al. (2009) compared NASA TLX scores in a Segmented Continuous Descent Approach and a Low Drag Low Power Approach. There was a significantly lower task load when the former type approach was performed for a second time than in the latter type approach. Moroney et al. (1993) reported that NASA TLX ratings were not significantly affected by the workload in a previous simulated flight. Workload was manipulated by crosswind level being simulated.

Boehm-Davis et al. (2010) reported significantly higher workload ratings for 24 air transport pilots using Data Comm than using voice. The data were collected in a simulator. In another simulator study, Bustamante et al. (2005) reported significantly higher time pressure ratings at 20 nautical mile (nm) to weather than 160 nm to weather. Their participants were 24 commercial airline pilots using a desk-top flight simulator. Baker et al. (2012) reported no significant difference in workload between voice and data communications for pilots using a low-fidelity flight simulator.

Nataupsky and Abbott (1987) successfully applied NASA TLX to a multi-task environment. Tsang and Johnson (1989) reported reliable increases in NASA TLX ratings when target-acquisition and engine-failure tasks were added to the primary flight task. Selcon et al. (1991) concluded from pilot ratings of an air combat flight simulation that NASA TLX was sensitive to difficulty but not the pilot experience.

Aretz et al. (1995) reported that the number of concurrent tasks had the greatest impact on NASA TLX ratings followed by participant's flight experience. Their data were collected in a fixed-base simulator flown by 15 U.S. Air Force Academy cadets with 0 to 15.9 flight hours.

Heers and Casper (1998) reported higher workload ratings without automatic terrain avoidance, missile warning receiver, and laser-guided rockets than with these advanced technologies. The data were collected in a Scout Helicopter simulator. The participants were eight U.S. Army helicopter pilots.

Air Traffic Control. On the Air Traffic Control side of flight, Pierce (2012) reported significant differences in workload between auditory shadowing and list memory conditions. Strybel et al. (2016) used NASA TLX to compare the workload of retired Air Traffic Controllers who used four different separation assurance and spacing concepts in enroute and transitional sectors. The author reported that workload was lowest when both functions were automated.

Metzger and Parasuraman (2005) reported significant differences in NASA TLX as a function of decision aid reliability in an air traffic scenario. Also in the Air Traffic domain, Vu et al. (2009) reported significantly higher workload ratings in high density air traffic scenarios than in low density scenarios. Willems and Heiney (2002) reported a significant main effect of task load and a significant interaction of task load with Air Traffic Controller position (data or radar controller). The researchers were evaluating no automation, limited automation, and full automation for enroute Air Traffic Control.

Durso et al. (1999) reported that the subscales were highly positively correlated except for the performance subscale in a civilian Air Traffic Control task.

Automobiles. Jeon and Zhang (2013) used NASA TLX to evaluate the effect of sadness, anger, and neutral emotions during a simulated driving experiment. Participants with induced anger reported higher workload than participants with neutral emotions. In another simulated driving task, Kennedy and Bliss (2013) reported higher mental demand on the NASA TLX for participants who detected a "no left turn" sign than those who failed to detect the sign.

Jordan and Johnson (1993) concluded from an on-road evaluation of a car stereo that NASA TLX was a useful measure of mental workload. Chrysler et al. (2010) used NASA TLX to measure driver workload on an open road and on a test track associated with high speed (60 versus 85 mph) driving. Szczerba et al. (2015) used NASA TLX to compare the workload among three automotive navigation systems: visual only, visual and auditory, or visual and haptic. They reported no significant differences among the systems.

Shah et al. (2015) reported no significant difference between auditory and tactile warnings on simulated driving.

Alm et al. (2006) used NASA TLX to assess night vision systems in passenger vehicles. NASA TLX ratings of mental demands and effort were significantly lower in the presence of the night vision system than when the system was absent. Jansen et al. (2016) concluded from a simulated driving study that there may be a learning effect associated with the NASA TLX.

NASA TLX has been used in evaluations of other vehicles besides aircraft and automobiles. For example, Mendel et al. (2011) used NASA TLX to assess workload in a back hoe with in-vehicle electronics.

Unmanned Systems. In an evaluation of Unmanned Aerial System (UAS) performance, Fern et al. (2012) used NASA TLX to compare workload with or without a Cockpit Situation Display and in high or low traffic density. There were no significant effects of display or traffic density. Helton et al. (2015) used NASA TLX to estimate team workload in a UAV simulation.

Minkov and Oron-Gilad (2009) reported significant differences in unweighted NASA TLX scores among Unmanned Aerial Vehicle (UAV) displays. In a similar study, Lemmers et al. (2004) used NASA TLX to evaluate a concept of operations for UAVs in civilian airspace in Europe.

Sellers et al. (2012) used MRQ to evaluate three levels of autonomy (Management by Consent, Management by Exception, and Full Autonomy) during the control of unmanned ground vehicles. They reported significantly higher workload in the first two levels of autonomy than in the third level.

Chen et al. (2010) reported no significant differences between 2D and 3D stereoscopic displays for robot teleoperation. In a U.S. Army study, Wright et al. (2016) compared NASA TLX measures of university students controlling a simulated robotic convoy. Unexpectedly, workload increased as access to an intelligent agent increased. In a comparison of levels of automation and number of simultaneously controlled remotely operated vehicles, Ruff et al. (2000) reported a significant interaction with the lowest NASA TLX score for manual control of one vehicle but highest for manual control for four vehicles. The automation modes were management by consent and management by exception. In a mixed human/robot experiment, Chen and Barnes (2012) reported significantly higher workload when controlling eight compared to controlling four robots. Riley and Strater (2006) used NASA TLX to compare four robot control modes and reported significant differences in workload.

NASA TLX has been applied to other environments. Hill et al. (1992) reported that the NASA TLX was sensitive to different levels of workload and high in user acceptance. Their participants were Army operators.

Fincannon et al. (2009a) reported lower workload with audio rather than text communication in teams of operators. Fincannon et al. (2009b) used NASA TLX to measure workload of teams of Unmanned Aerial Vehicle (UAV) and Unmanned Ground Vehicle (UGV) operators. Requesting support from the UAV operator did not increase workload but providing that support did.

Fouse et al. (2012) used NASA TLX to evaluate the effects of team controlling heterogeneous versus homogenous sets of four versus eight unmanned underwater vehicles. Teams with homogenous fleets of vehicles reported significantly less frustration but more effort than teams with heterogeneous fleets. Strang et al. (2012) reported that a cross trained team reported significantly higher workload than the control team. The task was simulated air battle management.

Nuclear Power Plants. Mercado et al. (2014) reported no significant effect of type of task performed in a nuclear power plant simulator (checking, detection, response implementation, and communication) on the global NASA TLX score but there was a significant effect on the frustration scale. Detection was rated significantly more frustrating than the other three tasks. Cosenzo et al. (2010) reported significant differences in physical demand and frustration scale among three levels of automated control of an unmanned ground vehicle (manual, adaptive, and static) but no effect on overall workload.

Laboratories. In a vigilance study, Claypoole et al. (2016) reported significantly higher workload when there was a passive supervisor than an active observer. Manzey et al. (2009) also used NASA TLX in a vigilance task and reported significantly higher workload ratings at night than day as well as without an automated decision aid as with it. In another vigilance task, Funke et al. (2016) reported no significant difference between paired observers working independently and observers working alone.

In an unusual application, Phillips and Madhavan (2012) reported higher workload scores for luggage screening with 9 second exposure than with 3 second exposure or continuous scrolling. Weber et al. (2013) reported no significant difference in workload as a function of system response time on a spam versus no spam email discrimination task. Note the version of the NASA TLX used was translated to German.

Kim et al. (2016) reported significantly higher workload as the complexity in a monitoring task increased. Dember et al. (1993) used the NASA TLX to measure workload on a visual vigilance task performed for 10, 20, 30, 40, or 50 minutes and at either an easy or difficult level of discrimination. They reported a linear increase in workload with time and a decrease with salience. Seidelman et al. (2012), using just the NASA TLX scales, reported no significant difference in workload in a bead transfer manual task between requiring transfer using a two or a four color pattern although there had been a significant difference between the two in a secondary time estimation task. Matthews et al. (2015) reported a significant effect size of 0.590 when comparing NASA TLX in scenarios with single versus dual change detection or single versus dual threat detection.

Dong and Hayes (2011) used NASA TLX to assess workload differences between two decision support systems. One presented relative uncertainty associated with an alternative (19 university students). The other presented relative certainty (14 university students). The former group reported significantly higher workload.

Fraulini et al. (2017) reported no significant differences in the global workload score comparing practice versus no practice of vigilance training and knowledge of results versus no knowledge of results. There was a significant difference on the frustration scale, however. No practice was associated with higher frustration scores than the practice condition. Temple et al. (2000) used NASA TLX to verify the high workload induced by a 12-minute computerized vigilance task. Szalma et al. (2004) found no significant effects on the global workload score of the NASA TLX for modality (auditory versus visual task) or time (four 10-minute vigilance periods).

In an auditory interference study, Ferraro et al. (2017) reported the expected significant difference in NASA TLX between performing with (higher) and without (lower) a communication secondary task. Kortum et al. (2010) reported a significant increase in NASA TLX score as duration of an Auditory Progress Bar (such as used during telephone on hold conditions) increased. There was no difference between type of auditory signal played (sine, cello, electronic), however. NASA TLX scores significantly increased as ambient noise increased (Becker et al., 1995). Shaw et al. (2012) reported no significant difference in NASA TLX overall workload score between monoaural radio and spatial radio command and control systems.

Hancock and Caird (1993) reported a significant increase in the overall workload rating scale of the NASA TLX as shrink rate of a target decreased. The highest ratings were on paths with four steps rather than 2, 8, or 16 steps from cursor to target. Isard and Szalma (2015) reported a significantly higher workload using NASA TLX when targets moved at 2 meters/second than at 1 meter per second. The same result occurred for the effort, performance, and temporal demand subscales. No effect on mental demand and an interaction with speed (1 versus 2 meters per second) and choice (easy versus hard task) on physical demand. Higher physical demand for 1 m/s with choice than those in no choice condition.

Bowers et al. (2014) also reported no significant differences in NASA TLX scores. Their participants performed the Air Force Multi-Attribute Task Battery at varying levels of event rate. Rodriguez Paras et al. (2015) reported a significant effect of a workload manipulation of varying either the difficulty or the number of tasks to be performed on the Multi-Attribute Task Battery-II.

Brill et al. (2009) reported higher workload with a visual detection task when a secondary task was added (visual counting, auditory counting, or tactile counting). Finomore et al. (2006) reported significant increase in NASA TLX for detecting the absence of a feature in a target than for detecting the presence of a feature in a target. Finomore et al. (2010) reported significant differences in NASA TLX scores for difficulty (easy versus hard) and communication format (radio, 3D audio, chat, and multi-modal). Jones and Endsley (2000) reported significant differences in NASA-TLX scores between war and peace scenarios in a military command and control experiment but no significant differences in response times to a secondary task in the same experiment.

Teo and Szalma (2010) reported significant effects of the number of visual displays monitored (1, 2, 4, or 8) on all six subscales of the NASA TLX, specifically as task demand increased the workload subscale rating increased. Also, event rate (8, 12, 16, or 20 events per minute) had a significant effect but only on two subscales – temporal demand and frustration. Jerome et al. (2006) used NASA TLX to assess augmented reality cues. They reported a significant decrease in workload with the presence of haptic cues. Liu (1996) asked participants to perform a pursuit tracking task with a decision-making task. Visual scanning resulted in increased workload as measured by the NASA TLX.

Riggs et al. (2014) used NASA TLX to compare navigational tools in a virtual environment. Teo and Szalma (2011) reported a significant effect of sensory (signals same or different size) versus cognition (signals same or different value) condition with the NASA TLX ratings being significantly higher in the former than the latter. There was also a significant increase in workload of monitoring 4 displays rather than 1 or 2. In a training study, Teo et al. (2013) reported no significant differences in workload between two training groups – one which received knowledge of results and one that did not.

Sawyer et al. (2014) reported that mental workload scale was rated high during a cyber security task. In the cyber threat domain, Giacobe (2013) reported no significant effects of text versus graphic presentation of information on NASA TLX scores. There was, however, a significant difference in the temporal demand scale between novice and experienced analysts.

Bowden and Rusnock (2016) reported significant differences in workload between varying strategies used to complete dual-task performance. In an unusual study, Dillard et al. (2015) reported no significant differences in NASA TLX between a group told that a task they performed for 30 minutes would last 15 minutes and a second group told it would last 60 minutes.

Satterfield et al. (2012) used NASA TLX to evaluate automation failures. There were significant differences between the control condition (no variation in the number of incoming threats) from both the single transition (one occurrence of an increase in threats) and the dual transition (two occurrences of increases in threats) though not between the single and dual conditions. Endsley and Kaber (1999) used NASA TLX to measure workload associated with varying levels of automation: manual control, action support, batch processing, shared control, decision support, blended decision making, rigid system, automated decision making, supervisory control, and full automation. Blended decision making, automated decision making, supervisory control, and full automation had significantly lower workload. Their participants were 30 undergraduate students. Harris et al. (1995) reported significantly higher ratings on five (mental demand, temporal demand, effort, frustration, and physical demand) of the six NASA-TLX scales for manual than automatic tracking.

In a gaming application, Kent et al. (2012) reported higher workload in the vocal input condition than in the manual input condition. In another modality study, Scerra and Brill (2012) reported significantly higher workload in dual-task performance of two tactile tasks than with mixed tactile and auditory task, mixed tactile and visual tasks, or single task. Crandall and Chaparro (2012) reported significantly greater mental workload and significantly greater physical demand using a touch screen to text while driving than using a physical keyboard.

In an unusual application of the NASA TLX, Szalma and Teo (2010) reported that individuals higher in neuroticism (as measured by N scale from the International Personality Item Pool) reported higher levels of frustration (one of the subscales on the NASA TLX) than individuals lower in neuroticism. In another personality trait study, Panganiban et al. (2011) measured the correlation between NASA TLX scale scores and State-Trait Personality Inventory (STPI) anxiety assessment scores. There were significant positive correlations between STPI first neutral block scores and NASA TLX physical demand and frustration scores as well as between STPI anxious block scores and NASA TLX mental demand, effort, and frustration scales. There were no significant correlations with STPI second neutral block and NASA TLX subscale scores. The participants were 46 individuals from local universities. The task was a simulated air defense task.

Vidulich and Pandit (1987) reported only three significant correlations between NASA TLX and seven personality tests (Jenkins Activity Survey, Rotter's Locus of Control, Cognitive Failures Questionnaire, Cognitive Interference Questionnaire, Thought Occurrence Questionnaire, California Q-Sort, and the Myers-Briggs Type Indicator): the speed scale of the Jenkins Activity Survey and the physical demand scale of the NASA TLX ($r = -0.23$), Locus of control and physical demand ($r = +0.21$), and finally locus of control and effort ($r = +0.23$). In a more recent study, Szalma (2002) reported no effect of stress coping strategy on NASA TLX after completing a 24-minute vigilance task. The participants were 48 male and 48 female undergraduate students.

In another atypical application, Fraune et al. (2013) used NASA TLX to compare the workload associated with creating a password using a picture memory aid or logging in. There were significant increases in physical demand during creation rather than logging in. In another unusual application, Newlin-Canzone et al. (2011) reported significantly higher workload in active (played the part of the interviewee) rather than passive observations of an interview.

Healthcare. In an evaluation of 2D and 3D displays for minimally invasive surgery, Sublette et al. (2010) reported no significant difference between the two types of displays. Lowndes et al. (2015) used NASA TLX to compare workload associated with differing laparoscopic surgical techniques. Warvel and Scerbo (2015) measured workload associated with varying camera angles also during laparoscopic surgery. Yu et al. (2015) used NASA TLX to measure workload associated with laparoscopic skills training.

Sublette et al. (2010) compared prospective and retrospective ratings for three psychomotor tasks. They reported a significant effect on the mental demand and the performance subscales on the NASA TLX such that the tasks rated lowest prospectively were rated most difficult retrospectively. The following year, Sublette et al. (2011) reported increased workload with increased number of surgical targets.

Levin et al. (2006) collected NASA TLX data from emergency room physicians after 180 minutes of tasking in an actual emergency room. High workloads were associated with exchanging patient information, directing patient care, completing phone calls and consults, and charting. The highest workload dimension was temporal demand. Lio et al. (2006) reported significant increases in NASA TLX as the precision requirements in a laparoscopic task increased. Grant et al. (2009) reported that NASA TLX was sensitive to blocks of time (workload decreased over block) and type of surgical task.

Luz et al. (2010) used NASA TLX after a simulated Mastoidectomy to evaluate the effectiveness of image guided navigation during surgery. They reported that surgical students rated workload significantly lower during the procedure manually than during that with the imagery system. Also in the medical domain, Mosaly et al. (2011) used NASA TLX to evaluate tasks performed by radiation oncology physicists during radiation treatment planning and delivery. In another medical application, McCrory et al. (2012) reported significantly lower overall and effort workload for a new intraoral mask than for a bag-valve mask.

In an unusual application of the NASA TLX, Kuehn et al. (2013) reported that participants with motor control disability rated touch screen operation as having higher workload than participants without a motor control disability.

Reliability. Battiste and Bortolussi (1988) reported significant workload effects as well as a test-retest correlation of +0.769. Corwin et al. (1989) reported that NASA TLX was a valid and reliable measure of workload.

Comparison to Other Workload Measures. Vidulich and Tsang (1985) compared the SWAT and the NASA TLX. They stated that the collection of ratings is simpler with SWAT. However, the SWAT card sort is more tedious and time consuming. Battiste and Bortolussi (1988) reported no significant correlation between SWAT and NASA TLX in a simulated B-727 flight. Hancock (1996) stated that NASA TLX and SWAT "were essentially equivalent in terms of their sensitivity to task manipulations." The task was tracking.

Jordan et al. (1995) reported the patterns of results for NASA-TLX and the Prediction of Performance (POP) workload measure was the same with the NASA-TLX having significant differences in all levels of workload conditions on a desk-top flight simulator. Tsang and Johnson (1987) reported good correlations between NASA TLX and a one-dimensional workload scale. Vidulich and Tsang (1987) replicated the Tsang and Johnson finding as well as reported a good correlation between NASA TLX and the Analytical

Hierarchy Process. Leggatt and Noyes (1997) reported that there was no significant difference in NASA TLX workload ratings between others and own ratings of a participant's workload. There was, however, an interaction in that subordinates rated the leader's workload higher than the leader rated his or her own workload. The participants were armored fighting vehicle drivers. Riley et al. (1994) compared results from NASA TLX with 22 methods of measuring workload. All methods gave approximately the same results for airline pilots completing Line Oriented Flight Training exercises. Windell et al. (2006) compared NASA TLX scores with a Short Subjective Instrument (SSI) (i.e., a single question for overall workload). The SSI showed significant differences between modules designed to vary in workload while the NASA TLX mental demands scale ratings did not.

Chin et al. (2004) described an adaptation of NASA TLX designed for driving, the Driving Activity Load Index (DALI). DALI factor, rated from low to high, are effort of attention, visual demand, auditory demand, tactile demand, temporal demand, and interference. DALI has been applied (Bashiri and Mann, 2013).

Single versus Multiple Estimates. Hendy et al. (1993) examined one-dimensional and multi-dimensional measures of workload in a series of four experiments (low-level helicopter operations, peripheral version display evaluation, flight simulator fidelity, and aircraft landing task). They concluded that if an overall measure of workload is required, then a univariate measure is as sensitive as an estimate derived from multivariate data. If a univariate measure is not available then a simple unweighted additive method can be used to combine ratings into an overall workload estimate. Peterson and Kozhokar (2017) reported no significant difference in overall NASA TLX score when the most challenging task was presented last. However, mental demand was rated significantly higher when the most challenging of the three tasks was performed last.

Byers et al. (1989) suggested using raw NASA TLX scores. Moroney et al. (1992) reported that the pre-rating weighting scheme is unnecessary since the correlation between weighted and unweighted scores was +0.94. Further, delays of 15 minutes did not affect the workload ratings; delays of 48 hours, however, did. After 48 hours, ratings no longer discriminate workload conditions. Moroney et al. (1995) concluded from a review of relevant studies that 15-minute delays do not affect NASA TLX.

Svensson et al. (1997) reported the reliability of the NASA TLX to be +0.77 among 18 pilots flying simulated low-level, high-speed missions. The correlation with the Bedford Workload Scale was +0.826 and with the SWAT was +0.735. Finomore et al. (2009) compared NASA TLX and Multiple Resources Questionnaire ratings in multi- and single-task conditions. Only the Multiple Resources Questionnaire showed greater workload in the multi-task environment.

In another comparison of workload measures, Rubio et al. (2004) compared workload estimates of NASA-TLX with those of SWAT and Workload Profile

(WP). Their participants were 36 psychology students who performed the Sternberg Memory Searching Task and a tracking task under single and dual-task conditions. There was no significant effect of memory set size in the Sternberg Memory Searching Task on the NASA-TLX ratings. Nor was the interaction of memory set size and tracking task path width in the dual-task condition significant. However, all other conditions had significant effects on NASA-TLX.

Loft et al. (2015) compared SPAM, SAGAT, ATWIT, NASA TLX, and SART ratings from 117 undergraduates performing three submarine tasks: contact classification, closest point of approach, and emergency surface. SPAM did not significantly correlate with SART but did with ATWIT and NASA TLX.

NASA TLX has been adapted to include team workload measures (Helton et al., 2014; Sellers et al., 2014). These authors concluded that workload measures vary between participants and within participants.

In an unusual study, Hale and Long (2017) used the NASA TLX to assess how well an observer could subjectively rate the mental workload of a participant. The observer's subjective scores underrated the participant's workload ratings and the same occurred in reverse.

Data requirements – Use of the TLX requires two steps. First, participants rate each task performed on each of the six subscales. Hart suggests that participants should practice using the rating scales in a training session. Second, participants must perform 15 pairwise comparisons of six workload scales. The number of times each scale is rated as contributing more to the workload of a task is used as the weight for that scale. Separate weights should be derived for diverse tasks; the same weights can be used for similar tasks. Note that a set of PC compatible programs has been written to gather ratings and weights and to compute the weighted workload scores. The programs are available from the Human Factors Division at NASA Ames Research Center, Moffett Field, California. NASA TLX can also be accessed online at http://NASATLX.com (Sharek, 2011).

Thresholds – Knapp and Hall (1990) used NASA TLX to evaluate a highly automated communication system. Using 40 as a high workload threshold, the system was judged to impose high workload and difficult cognitive effort on operators. Sturrock and Fairburn (2005) defined red line values:

Development/risk reduction workload assessments

    0–60 acceptable

    61–80 investigate further

    81–100 unacceptable probable design change

Qualification workload assessments

    0–80 acceptable

    81–100 investigate design change

Sources

Alexander, A.L., Stelzer, E.M., Kim, S.H., Kaber, D.B., and Prinzel, L.J. Data and knowledge as predictors of perceptions of display clutter, subjective workload and pilot performance. Proceedings of the Human Factors and Ergonomics Society 53rd Annual Meeting, 21–25, 2009.

Alm, T., Kovordanyi, R., and Ohlsson, K. Continuous versus situation-dependent night vision presentation in automotive applications. Proceedings of the Human Factors and Ergonomics Society 50th Annual Meeting, 2033–2037, 2006.

Aretz, A.J., Shacklett, S.F., Acquaro, P.L., and Miller, D. The prediction of pilot subjective workload ratings. Proceedings of the Human Factors and Ergonomics Society 39th Annual Meeting, 94–97, 1995.

Baker, K.M., DiMare, S.K., Nelson, E.T., and Boehm-Davis, D.A. Effect of data communications on pilot situation awareness, decision making, and workload. Proceedings of the Human Factors and Ergonomics Society 56th Annual Meeting, 1787–1788, 2012.

Bashiri, B., and Mann, D.D. Drivers' mental workload in agricultural semi-autonomous vehicles. Proceedings of the Human Factors and Ergonomics Society 57th Annual Meeting, 1795–1799, 2013.

Battiste, V., and Bortolussi, M.R. Transport pilot workload: A comparison of two objective techniques. Proceedings of the Human Factors Society 32nd Annual Meeting, 150–154, 1988.

Becker, A.B., Warm, J.S., Dember, W.N., and Hancock, P.A. Effects of jet engine noise and performance feedback on perceived workload in a monitoring task. *International Journal of Aviation Psychology* 5(1): 49–62, 1995.

Bittner, A.C., Byers, J.C., Hill, S.G., Zaklad, A.L., and Christ, R.E. Generic workload ratings of a mobile air defense system. Proceedings of the Human Factors Society 33rd Annual Meeting, 1476–1480, 1989.

Boehm-Davis, D.A., Gee, S.K., Baker, K., and Medina-Mora, M. Effect of party line loss and delivery format on crew performance and workload. Proceedings of the Human Factors and Ergonomics Society 54th Annual Meeting, 126–130, 2010.

Bowden, J.R., and Rusnock, C.F. Influences of task management strategy on performance and workload for supervisory control. Proceedings of the Human Factors and Ergonomics Society 60th Annual Meeting, 855–859, 2016.

Bowers, M.A., Christensen, J.C., and Eggemeier, F.T. The effects of workload transitions in a multitasking environment. Proceedings of the Human Factors and Ergonomics Society 58th Annual Meeting, 220–224, 2014.

Brill, J.C., Mouloua, M., and Hendricks, S.D. Compensatory strategies for managing the workload demands of a multimodal reserve capacity task. Proceedings of the Human Factors and Ergonomics Society 53rd Annual Meeting, 1151–1155, 2009.

Brown, R.D., and Galster, S.M. Effects of reliable and unreliable automation on subjective measures of mental workload, situation awareness, trust and confidence in a dynamic flight task. Proceedings of the Human Factors and Ergonomics Society 48th Annual Meeting, 147–151, 2004.

Bustamante, E.A., Fallon, C.K., Bliss, J.P., Bailey, W.R., and Anderson, B.L. Pilots' workload, situation awareness, and trust during weather events as a function of time pressure, role assignment, pilots' rank, weather display, and weather system. *International Journal of Applied Aviation Studies* 5(2): 348–368, 2005.

Byers, J.C., Bittner, A.C., and Hill, S.G. Traditional and raw Task Load Index (TLX) correlations: Are paired comparisons necessary? In Advances in Industrial Ergonomics and Safety (pp. 481–485). London: Taylor & Francis Group, 1989.

Byers, J.C., Bittner, A.C., Hill, S.G., Zaklad, A.L., and Christ, R.E. Workload assessment of a remotely piloted vehicle (RPV) system. Proceedings of the Human Factors Society 32nd Annual Meeting, 1145–1149, 1988.

Casner, S.M. Perceived vs. measured effects of advanced cockpit systems on pilot workload and error: Are pilots' beliefs misaligned with reality? *Applied Ergonomics* 40: 448–456, 2009.

Chen, J.Y.C., and Barnes, M.J. Supervisory control of multiple robots: Effects of imperfect automation and individual differences. *Human Factors* 54(2): 157–174; 2012.

Chen, J.Y.C., Oden, R.N.V., Kenny, C., and Merritt, J.O. Stereoscopic displays for robot teleoperation and simulated driving. Proceedings of the Human Factors and Ergonomics Society 54th Annual Meeting, 1488–1492, 2010.

Chin, E., Nathan, F., Pauzie, A., Manzano, J., Nodari, E., Cherri, C., Rambaldini, A., Toffetti, A., and Marchitto, M. Subjective assessment methods for workload. Information Society Technologies (IST) Programme Adaptive Integrated Driver-vehicle Interface (AIDE)(ISR-1-507674-IP). Gothenburg, Sweden: Information Society Technologies Program, March 2004.

Chrysler, S.T., Funkhouser, D., Fitzpatrick, K., and Brewer, M. Driving performance and driver workload at high speeds: Results from on-road and test track studies. Proceedings of the Human Factors and Ergonomics Society 54th Annual Meeting, 2071–2075, 2010.

Claypoole, V.L., Dewar, A.R., Fraulini, N.W., and Szalma, J.L. Effects of social facilitation on perceived workload, subjective stress, and vigilance-related anxiety. Proceedings of the Human Factors and Ergonomics Society 60th Annual Meeting, 1168–1172, 2016.

Corwin, W.H., Sandry-Garza, D.L., Biferno, M.H., Boucek, G.P., Logan, A.L., Jonsson, J.E., and Metalis, S.A. *Assessment of Crew Workload Measurement Methods, Techniques, and Procedures. Volume I-Process, Methods, and Results (WRDC-TR-89-7006).* Wright-Patterson Air Force Base, OH, 1989.

Cosenzo, K., Chen, J., Reinerman-Jones, L., Barnes, M., and Nicholson, D. Adaptive automation effects on operator performance during a reconnaissance mission with an unmanned ground vehicle. Proceedings of the Human Factors and Ergonomics Society 54th Annual Meeting, 2135–2139, 2010.

Crandall, J.M., and Chaparro, A. Driver distraction: Effects of text entry methods on driving performance. Proceedings of the Human Factors and Ergonomics Society 56th Annual Meeting, 1693–1697, 2012.

Dember, W.N., Warm, J.S., Nelson, W.T., Simons, K.G., and Hancock, P.A. The rate of gain of perceived workload in sustained operations. Proceedings of the Human Factors and Ergonomics Society 37th Annual Meeting, 1388–1392, 1993.

Dillard, M.B., Warm, J.S., Funke, G.J., Vidulich, M.A., Nelson, W.T., Eggemeier, T.F., and Funke, M.E. Vigilance: Hard work even if time flies. Proceedings of the Human Factors and Ergonomics Society 57th Annual Meeting, 1114–1118, 2015.

Dong, X., and Hayes, C. The impact of uncertainty visualization on team decision making. *Proceedings of the Human Factors and Ergonomics Society 55th Annual Meeting*, 257–261, 2011.

Durso, F.T., Hackworth, C.A., Truitt, T.R., Crutchfield, J., Nikolic, D., and Manning, C.A. *Situation Awareness as a Predictor of Performance in En Route Air Traffic Controllers (DOT/FAA/AM-99/3)*. Washington, DC: Office of Aviation Medicine, January 1999.

Elmenhorst, E., Vejvoda, M., Maass, H., Wenzel, J., Plath, G., Schubert, E., and Basner, M. Pilot workload during approaches: Comparison of simulated standard and noise-abatement profiles. *Aviation, Space, and Environmental Medicine* 80: 364–370, 2009.

Endsley, M.R., and Kaber, D.B. Level of automation effects on performance, situation awareness, and workload in a dynamic control task. *Ergonomics* 42(3): 462–492, 1999.

Ferraro, J., Christy, N., and Mouloua, M. Impact of auditory interference on automated task monitoring and workload. *Proceedings of the Human Factors and Ergonomics Society Annual Meeting*, 1136–1140, 2017.

Fern, L., Flaherty, S.R., Shively, R.J., and Turpin, T.S. Airspace deconfliction for UAS operations. *16th International Symposium on Aviation Psychology*. 451–456, 2011.

Fern, L., Kenny, C.A., Shively, R.J., and Johnson, W. UAS integration into the NAS: An examination of baseline compliance in the current airspace system. *Proceedings of the Human Factors and Ergonomics Society 56th Annual Meeting*, 41–45, 2012.

Fincannon, T.D., Evans, A.W., Jentsch, F., Phillips, E., and Keebler, J. Effects of sharing control of unmanned vehicles on backup behavior and workload in distributed operator teams. *Proceedings of the Human Factors and Ergonomics Society 53rd Annual Meeting*, 1300–1303, 2009a.

Fincannon, T.D., Evans, A.W., Phillips, E., Jentsch, F., and Keebler, J. The influence of team size and communication modality on team effectiveness with unmanned systems. *Proceedings of the Human Factors and Ergonomics Society 53rd Annual Meeting*, 419–423, 2009a.

Finomore, V.S., Shaw, T.H., Warm, J.S., Matthews, G., Weldon, D., and Boles, D.B. On the workload of vigilance: Comparison of the NASA-TLX and the MRQ. *Proceedings of the Human Factors and Ergonomics Society 53rd Annual Meeting*, 1057–1061, 2009.

Finomore, V., Popik, D., Castle, C., and Dallman, R. Effects of a network-centric multimodal communication tool on a communication monitoring task. *Proceedings of the Human Factors and Ergonomics Society 54th Annual Meeting*, 2125–2129, 2010.

Finomore, V.S., Warm, J.S., Matthews, G., Riley, M.A., Dember, W.N., Shaw, T.H., Ungar, N.R., and Scerbo, M.W. Measuring the workload of sustained attention. *Proceedings of the Human Factors and Ergonomics Society 50th Annual Meeting*, 1614–1618, 2006.

Fouse, S., Champion, M., and Cooke, N.J. The effects of vehicle number and function on performance and workload in human-robot teaming. *Proceedings of the Human Factors and Ergonomics Society 56th Annual Meeting*, 398–402, 2012.

Fraulini, N.W., Fistel, A.L., Perez, M.A., Perez, T.L., and Szalma, J.L. Examining the effects of a novel training paradigm for vigilance on mental workload and stress. Proceedings of the Human Factors and Ergonomics Society Annual Meeting, 1504–1508, 2017.

Fraune, M.R., Juang, K.A., Greenstein, J.S., Madathil, K.C., and Koikkara, R. Employing user-created pictures to enhance the recall of system-generated mnemonic phrases and the security of passwords. Proceedings of the Human Factors and Ergonomics Society 57th Annual Meeting, 419–423, 2013.

Funke, G.J., Warm, J.S., Baldwin, C.L., Garcia, A., Funke, M.E., Dillard, M.B., Finomore, V.S., Mathews, G., and Greenlee, E.T. The independence and interdependence of coacting observers in regard to performance efficiency, workload, and stress in a vigilance task. *Human Factors* 58(6): 915–926, 2016.

Giacobe, N.A. A picture is worth a thousand words. Proceedings of the Human Factors and Ergonomics Society 57th Annual Meeting, 172–176, 2013.

Grant, R.C., Carswell, C.M., Lio, C.H., Seales, B., and Clarke, D. Verbal time production as a secondary task: Which metrics and target intervals are most sensitive to workload for fine motor laparoscopic training tasks? Proceedings of the Human Factors and Ergonomics Society 53rd Annual Meeting, 1191–1195, 2009.

Grubb, P.L., Warm, J.S., Dember, W.N., and Berch, D.B. Effects of multiple-signal discrimination on vigilance performance and perceived workload. Proceedings of the Human Factors and Ergonomics Society 39th Annual Meeting, 1360–1364, 1995.

Hale, L.T., and Long, P.A. How accurately can an observer assess participant self-reported workload? Proceedings of the Human Factors and Ergonomics Society Annual Meeting, 1486–1487, 2017.

Hancock, P.A. Effects of control order, augmented feedback, input device, and practice on tracking performance and perceived workload. *Ergonomics* 39(9): 1146–1162, 1996.

Hancock, P.A., and Caird, J.K. Experimental evaluation of a model of mental workload. *Human Factors* 35(3): 413–419, 1993.

Hancock, P.A., William, G., Manning, C.M., and Miyake, S. Influence of task demand characteristics on workload and performance. *International Journal of Aviation Psychology* 5(1): 63–86, 1995.

Harris, W.C., Hancock, P.A., Arthur, E.J., and Caird, J.K. Performance, workload, and fatigue changes associated with automation. *International Journal of Aviation Psychology* 5(2): 169–185, 1995.

Hart, S.G., Hauser, J.R., and Lester, P.T. Inflight evaluation of four measures of pilot workload. Proceedings of the Human Factors Society 28th Annual Meeting, 945–949, 1984.

Hart, S.G., and Staveland, L.E. Development of NASA-TLX (Task Load Index): Results of empirical and theoretical research. http://stavelandhfe.com/images/NASA-TLX_paper.pdf, 1987.

Heers, S.T., and Casper, P.A. Subjective measurement assessment in a full mission scout-attack helicopter simulation. Proceedings of the Human Factors and Ergonomics Society 42nd Annual Meeting, 26–30, 1998.

Helton, W.S., Epling, S., de Joux, N., Funke, G.J., and Knott, B.A. Judgments of team workload and stress: A simulated Unmanned Aerial Vehicle case. Proceedings of the Human Factors and Ergonomics Society 59th Annual Meeting, 736–740, 2015.

Helton, W.S., Funke, G.J., and Knott, B.A. Measuring workload in collaborative contexts: Trait versus state perspectives. *Human Factors* 56(2): 322–332, 2014.

Hendy, K.C., Hamilton, K.M., and Landry, L.N. Measuring subjective workload: When is one scale better than many? *Human Factors* 35(4): 579–601, 1993.

Hill, S.G., Byers, J.C., Zaklad, A.L., and Christ, R.E. Subjective workload assessment during 48 continuous hours of LOS-F-H operations. Proceedings of the Human Factors Society 33rd Annual Meeting, 1129–1133, 1989.

Hill, S.G., Iavecchia, H.P., Byers, J.C., Bittner, A.C., Zaklad, A.L., and Christ, R.E. Comparison of four subjective workload rating scales. *Human Factors* 34: 429–439, 1992.

Hill, S.G., Zaklad, A.L., Bittner, A.C., Byers, J.C., and Christ, R.E. Workload assessment of a mobile air defense system. Proceedings of the Human Factors Society 32nd Annual Meeting, 1068–1072, 1988.

Isard, J.L., and Szalma, J.L. The effect of perceived choice on performance, workload, and stress. Proceedings of the Human Factors and Ergonomics Society, 1037–1041, 2015.

Jansen, R.J., Sawyer, B.D., van Egmond, R., de Ridder, H., and Hancock, P.A. Hysteresis in mental workload and task performance: The influence of demand transitions and task prioritization. *Human Factors* 58(8): 1143–1157, 2016.

Jeon, M., and Zhang, W. Sadder but wiser? Effects of negative emotions on risk perception, driving performance, and perceived workload. Proceedings of the Human Factors and Ergonomics Society 57th Annual Meeting, 1849–1853, 2013.

Jerome, C.J., Witner, B., and Mouloua, M. Attention orienting in augmented reality environments: Effects of multimodal cues. Proceedings of the Human Factors and Ergonomics Society 50th Annual Meeting, 2114–2118, 2006.

Jones, D.G., and Endsley, M.R. Can real-time probes provide a valid measure of situation awareness? Proceedings of the 1st hHuman Performance, Situation Awareness and Automation: User-Centered Design for the New Millennium, 245–250, 2000.

Jordan, C.S., Farmer, E.W., and Belyavin, A.J. The DRA Workload Scales (DRAWS): A validated workload assessment technique. Proceedings of the 8th International Symposium on Aviation Psychology, 1013–1018, 1995.

Jordan, P.W., and Johnson, G.L. Exploring mental workload via TLX: The case of operating a car stereo whilst driving. In A.G. Gale, I.D. Brown, C.M. Haslegrave, H.W. Kruysse, and S.P. Taylor (Eds.) *Vision in Vehicles – IV* (pp. 255–262). Amsterdam: North-Holland, 1993.

Keillor, J., Ellis, K., Craig, G., Rozovski, D., and Erdos, R. Studying collision avoidance by nearly colliding: A flight test evaluation. Proceedings of the Human Factors and Ergonomics Society 55th Annual Meeting, 41–45, 2011.

Kennedy, K.D., Stephens, C.L., Williams, R.A., and Schutte, P.C. Automation and inattentional blindness in a simulated flight task. Proceedings of the Human Factors and Ergonomics Society 58th Annual Meeting, 2058–2062, 2014.

Kennedy, K.K., and Bliss, J.P. Intentional blindness in a simulated driving task. Proceedings of the Human Factors and Ergonomics Society 57th Annual Meeting, 1899–1903, 2013.

Kent, T.M., Marraffino, M.D., Najle, M.B., Sinatra, A.M., and Sims, V.K. Effects of input modality and expertise on workload and video game performance. Proceedings of the Human Factors and Ergonomics Society 56th Annual Meeting, 1069–1073, 2012.

Kim, J.H., Yang, X., and Putri, M. Multitasking performance and workload during a continuous monitoring task. Proceedings of the Human Factors and Ergonomics Society 60th Annual Meeting, 665–669, 2016.

Knapp, B.G., and Hall, B.J. High performance concerns for the TRACKWOLF system (ARI Research Note 91–14). Alexandria, VA, 1990.

Kortum, P., Peres, S.C., and Stallman, K. Mental workload measures of auditory stimuli heard during periods of waiting. Proceedings of the Human Factors and Ergonomics Society 54th Annual Meeting, 1689–1693, 2010.

Kuehn, K.A., Chourasia, A.O., Wiegmann, D.A., and Sesto, M.E. Effects of orientation on workload during touchscreen operation among individuals with and without disabilities. Proceedings of the Human Factors and Ergonomics Society 57th Annual Meeting, 1580–1584, 2013.

Leggatt, A., and Noyes, J. Workload judgments: Self-assessment versus assessment of others. In D. Harris (Ed.) *Engineering Psychology and Cognitive Ergonomics Volume One Transportation Systems* (pp. 443–449). Aldershot, UK: Ashgate, 1997.

Lee, Y., and Liu, B. Inflight workload assessment: Comparison of subjective and physiological measurements. *Aviation, Space, and Environmental Medicine* 74(10): 1078–1084, 2003.

Lemmers, A., Valens, M., Beemster, T., Schmitt, D., and Klostermann, E. Unmanned aerial vehicle safety issues for civil operations (D4.1/WP4100 Report). USICO European Commission, April 9, 2004.

Levin, S., France, D.J., Hemphill, R., Jones, I., Chen, K.Y., Rickard, D., Makowski, R., and Aronsky, D. Tracking workload in the emergency department. *Human Factors* 48(3): 526–539, 2006.

Lio, C.H., Bailey, K., Carswell, C.M., Seales, W.B., Clarke, D., and Payton, G.M. Time estimation as a measure of mental workload during the training of laparoscopic skills. Proceedings of the Human Factors and Ergonomics Society 50th Annual Meeting, 1910–1913, 2006.

Liu, Y. Quantitative assessment of effects of visual scanning on concurrent task performance. *Ergonomics* 39(3): 382–399, 1996.

Loft, S., Bowden, V., Braithwaite, J., Morrell, D.B., Huf, S., and Durso, F.T. Situation awareness measures for simulated submarine track management. *Human Factors* 57(2): 298–310, 2015.

Lowndes, B., Abdelrahman, A., McCrory, B., and Hallbeck, S. A preliminary study of novice workload and performance during surgical simulation tasks for conventional versus single incision laparoscopic techniques. Proceedings of the Human Factors and Ergonomics Society 59th Annual Meeting, 498–502, 2015.

Luz, M., Mueller, S., Strauss, G., Dietz, A., Meixenberger, J., and Manzey, D. Automation in surgery: The impact of navigation-control assistance on the performance, workload and situation awareness of surgeons. Proceedings of the Human Factors and Ergonomics Society 54th Annual Meeting, 889–893, 2010.

Manzey, D., Reichenbach, J., and Onnasch, L. Human performance consequences of automated decision aids in states of fatigue. Proceedings of the Human Factors and Ergonomics Society Annual Meeting, 329–333, 2009.

Matthews, G., Reinerman-Jones, L.E., Barber, D.J., and Abich, J. The psychometrics of mental workload: Multiple measures are sensitive but divergent. *Human Factors* 57(1): 125–143, 2015.

McCrory, B., Lowndes, B.R., Thompson, D.L., Miller, E.E., Riggle, J.D., Wadman, M.C., and Hallbeck, M.S. Workload comparison of intraoral mask to standard mask ventilation using a cadaver model. Proceedings of the Human Factors and Ergonomics Society 56th Annual Meeting, 1728–1732, 2012.

Mendel, J., Pak, R., and Drum, J.E. Designing for consistency reduce workload in dual-task situations? Proceedings of the Human Factors and Ergonomics Society 55th Annual Meeting, 2000–2004, 2011.

Mercado, J.E., Reinerman-Jones, L., Barber, D., and Leis, R. Investigating workload measures in the nuclear domain. Proceedings of the Human Factors and Ergonomics Society 58th Annual Meeting, 205–209, 2014.

Metzger, U., and Parasuraman, R. Automation in future air traffic management: Effects of decision aid reliability on controller performance and mental workload. *Human Factors* 47(1): 35–49, 2005.

Minkov, Y., and Oron-Gilad, T. Display type effects in military operational tasks using UAV video images. Proceedings of the Human Factors and Ergonomics Society 53rd Annual Meeting, 71–75, 2009.

Moroney, W.F., Biers, D.W., and Eggemeier, F.T. Some measurement and methodological considerations in the application of subjective workload measurement techniques. *International Journal of Aviation Psychology* 5(1): 87–106, 1995.

Moroney, W.E., Biers, D.W., Eggemeier, F.T., and Mitchell, J.A. A comparison of two scoring procedures with the NASA Task Load Index in a simulated flight task. NAECON Proceedings, 734–740, 1992.

Moroney, W.F., Reising, J., Biers, D.W., and Eggemeier, D.W. The effect of previous level of workload on the NASA Task Load Index (TLX) in a simulated flight environment. Proceedings of the 7th International Symposium on Aviation Psychology, 882–890, 1993.

Mosaly, P.R., Mazur, L.M., Jackson, M., Chang, S.X., Deschesne Burkhardt, K., Jones, E.L., Xu, J., Rockwell, J., and Marks, L.B. Empirical evaluation of workload of the radiation oncology physicist during radiation treatment planning and delivery. Proceedings of the Human Factors and Ergonomics Society 55th Annual Meeting, 753–757, 2011.

Nataupsky, M., and Abbott, T.S. Comparison of workload measures on computer-generated primary flight displays. Proceedings of the Human Factors Society 31st Annual Meeting, 548–552, 1987.

Newlin-Canzone, E.T., Scerbo, M.W., Gliva-McConvey, G., and Wallace, A. Attentional and mental workload demands in nonverbal communication. Proceedings of the Human Factors and Ergonomics Society 55th Annual Meeting, 1190–1194, 2011.

Nygren, T.E. Psychometric properties of subjective workload measurement techniques: Implications for their use in the assessment of perceived mental workload. *Human Factors* 33 (1): 17–33, 1991.

Panganiban, A.R., Matthews, G., Funke, G., and Knott, B.A. Effects of anxiety on performance and workload in an air defense task. Proceedings of the Human Factors and Ergonomics Society 55th Annual Meeting, 909–913, 2011.

Peterson, D.A., and Kozhokar, D. Peak-end effects for subjective mental workload ratings. Proceedings of the Human Factors and Ergonomics Society Annual Meeting, 2052–2056, 2017.

Phillips, R.R., and Madhavan, P. The effect of simulation style on performance. Proceedings of the Human Factors and Ergonomics Society 56th Annual Meeting, 353–397, 2012.

Pierce, R.S. The effect of SPAM administration during a dynamic simulation. *Human Factors* 54(5): 838–848, 2012.

Riggs, A., Melloy, B.J., and Neyens, D.M. The effect of navigational tools and related experience on task performance in a virtual environment. Proceedings of the Human Factors and Ergonomics Society 58th Annual Meeting, 2378–2382, 2014.

Riley, J.M., and Strater, L.D. Effects of robot control mode on situational awareness and performance in a navigation task. Proceedings of the Human Factors and Ergonomics Society 50th Annual Meeting, 540–544, 2006.

Riley, V., Lyall, E., and Wiener, E. Analytic workload models for flight deck design and evaluation. Proceedings of the Human Factors and Ergonomics Society 38th Annual Meeting, 81–84, 1994.

Rodes, W., and Gugerty, L. Effects of electronic map displays and individual differences in ability on navigation performance. *Human Factors* 54(4): 589–599, 2012.

Rodriguez Paras, C., Yang, S., Tippey, K., and Ferris, T.K. Physiological indicators of the cognitive redline. Proceedings of the Human Factors and Ergonomics Society 59th Annual Meeting, 637–641, 2015.

Rubio, S., Diaz, E., Martin, J., and Puente, J.M. Evaluation of subjective mental workload: A comparison of SWAT, NASA-TLX, and Workload Profile Methods. *Applied Psychology: An International Review* 53(1): 61–86, 2004.

Ruff, H.A., Draper, M.H., and Narayanan, S. The effect of automation level and decision aid fidelity on the control of multiple remotely operated vehicles. Proceedings of the 1st Human Performance, Situation Awareness and Automation: User-Centered Design for the New Millennium, 70–75, 2000.

Satterfield, K., Ramirez, R., Shaw, T., and Parasuraman, R. Measuring workload during a dynamic supervisory control task using cerebral blood flow velocity and the NASA TLX. Proceedings of the Human Factors and Ergonomics Society 56th Annual Meeting, 163–167, 2012.

Sawyer, B.D., Finomore, V.S., Funke, G.J., Mancuso, V.F., Funke, M.E., Matthews, G., and Warm, J.S. Cyber vigilance: Effects of signal probability and event rate. Proceedings of the Human Factors and Ergonomics Society 58th Annual Meeting, 1771–1775, 2014.

Scerra, V.E., and Brill, J.C. Effect of task modality on dual-task performance, response time, and rating of operator workload. Proceedings of the Human Factors and Ergonomics Society 56th Annual Meeting, 1456–1460, 2012.

Seidelman, W., Carswell, C.M., Grant, R.C., Sublette, M., Lio, C.H., and Seales, B. Interval production as a secondary task workload measure: Consideration of primary task demands for interval selection. Proceedings of the Human Factors and Ergonomics Society 56th Annual Meeting, 1664–1668, 2012.

Selcon, S.J., Taylor, R.M., and Koritsas, E. Workload or situational awareness?: TLX vs. SART for aerospace systems design evaluation. Proceedings of the Human Factors Society 35th Annual Meeting, 62–66, 1991.

Sellers, B.C., Fincannon, T., and Jentsch, F. The effects of autonomy and cognitive abilities on workload and supervisory control of unmanned systems. Proceedings of the Human Factors and Ergonomics Society 56th Annual Meeting, 1039–1043, 2012.

Sellers, J., Helton, W.S., Naswall, K., Funke, G.L., and Knott, B.A. Development of the Team Workload Questionnaire. Proceedings of the Human Factors and Ergonomics Society 58th Annual Meeting, 989–993, 2014.

Shah, S.J., Bliss, J.P., Chancey, E.T., and Brill, J.C. Effects of alarm modality and alarm reliability on workload, trust, and driving performance. Proceedings of the Human Factors and Ergonomics Society 59th Annual Meeting, 1535–1539, 2015.

Sharek, D. A useable, online NASA-TLX tool. Proceedings of the Human Factors and Ergonomics Society 55th Annual Meeting, 1375–1379, 2011.

Shaw, T.H., Satterfield, K., Ramirez, R., and Finomore, V. A comparison of subjective and physiological workload assessment techniques during a 3-dimensional audio vigilance task. Proceedings of the Human Factors and Ergonomics Society 56th Annual Meeting, 1451–1455, 2012.

Stark, J.M., Comstock, J.R., Prinzel, L.J., Burdette, D.W., and Scerbo, M.W. A preliminary examination of situation awareness and pilot performance in a synthetic vision environment. Proceedings of the Human Factors and Ergonomics Society 45th Annual Meeting, 40–43, 2001.

Strang, A.J., Funke, G.J., Knott, B.A., Galster, S.M., and Russell, S.M. Effects of cross-training on team performance, communication, and workload in simulated air battle management. Proceedings of the Human Factors and Ergonomics Society 56th Annual Meeting, 1581–1585, 2012.

Strybel, T.Z., Vu, K.L., Chiappe, D.L., Morgan, C.A., Morales, G., and Battiste, V. Effects of NextGen concepts of operation for separation assurance and interval management on Air Traffic Controller situation awareness, workload, and performance. *International Journal of Aviation Psychology* 26(1–2): 1–14, 2016.

Sturrock, F., and Fairburn, C. Measuring pilot workload in single and multi-crew aircraft. Measuring pilot workload in a single and multi-crew aircraft. Contemporary Ergonomics 2005: Proceedings of the International Conference on Contemporary Ergonomics, 588–592, 2005.

Sublette, M., Carswell, C.M., Grant, R., Seidelman, G.W., Clarke, D., and Seales, W.B. Anticipating workload: Which facets of tack difficulty are easiest to predict? Proceedings of the Human Factors and Ergonomics Society 54th Annual Meeting, 1704–1708, 2010.

Sublette, M., Carswell, C.M., Han, Q., Grant, R., Lio, C.H., Lee, G., Field, M., Staley, D., Seales, W.B., and Clarke, D. Dual-view displays for minimally invasive surgery: Does the addition of a 3D global view decrease mental workload? Proceedings of the Human Factors and Ergonomics Society 54th Annual Meeting, 1581–1585, 2010.

Sublette, M., Carswell, C.M., Seidelman, W., Grant, R., Han, W., Field, M., Lio, C.H., Lee, G., Seales, W.B., and Clarke, D. Do operators take advantage of a secondary, global-perspective display when performing a simulated laparoscopic search task? Proceedings of the Human Factors and Ergonomics Society 55th Annual Meeting, 1626–1630, 2011.

Svensson, E., Angelborg-Thanderz, M., Sjoberg, L., and Olsson, S. Information complexity – Mental workload and performance in combat aircraft. *Ergonomics* 40(3): 362–380, 1997.

Szalma, J.L. Individual difference in the stress and workload of sustained attention. Proceedings of the Human Factors and Ergonomics Society 46th Annual Meeting, 1002–1006, 2002.

Szalma, J.L., and Teo, G.W.L. The joint effect of task characteristics and neuroticism on the performance, workload, and stress of signal detection. Proceedings of the Human Factors and Ergonomics Society 54th Annual Meeting, 1052–1056, 2010.

Szalma, J.L., Warm, J.S., Matthews, G., Dember, W.N., Weiler, E.M., Meier, A., and Eggemeier, F. T. Effects of sensory modality and task duration on performance, workload, and stress in sustained attention. *Human Factors* 46(2): 219–233, 2004.

Szczerba, J., Hersberger, R., and Mathieu, R. A wearable vibrotactile display for automotive route guidance: Evaluating usability, workload, performance and preference. Proceedings of the Human Factors and Ergonomics Society 59th Annual Meeting, 1027–1031, 2015.

Temple, J.G., Warm, J.S., Dember, W.N., Jones, K.S., LaGrange, C.M., and Matthews, G. The effects of signal salience and caffeine on performance, workload, and stress in an abbreviated vigilance task. *Human Factors* 42(2): 183–194, 2000.

Teo, G.W., Schmidt, T.N., Szalma, J.L., Hancock, G.M., and Hancock, P.A. The effects of feedback in vigilance training on performance, workload, stress and coping. Proceedings of the Human Factors and Ergonomics Society 57th Annual Meeting, 1119–1123, 2013.

Teo, G.W.L., and Szalma, J.L. The effect of spatial and temporal task characteristics on performance, workload, and stress. Proceedings of the Human Factors and Ergonomics Society 54th Annual Meeting, 1699–1703, 2010.

Teo, G., and Szalma, J.L. The effects of task type and source complexity on vigilance performance, workload, and stress. Proceedings of the Human Factors and Ergonomics Society 55th Annual Meeting, 1180–1184, 2011.

Tsang, P.S., and Johnson, W. Automation: Changes in cognitive demands and mental workload. Proceedings of the 4th Symposium on Aviation Psychology, 616–622, 1987.

Tsang, P.S., and Johnson, W.W. Cognitive demands in automation. *Aviation, Space, and Environmental Medicine* 60: 130–135, 1989.

Vidulich, M.A., and Bortolussi, M.R. Control configuration study. Proceedings of the American Helicopter Society National Specialist's Meeting: Automation Application for Rotorcraft, 20–29, 1988a.

Vidulich, M.A., and Bortolussi, M.R. Speech recognition in advanced rotorcraft: Using speech controls to reduce manual control overload. Proceedings of the National Specialists' Meeting Automation Applications for Rotorcraft, 20–30, 1988b.

Vidulich, M.A., and Pandit, P. Individual differences and subjective workload assessment: Comparing pilots to nonpilots. Proceedings of the International Symposium on Aviation Psychology, 630–636, 1987.

Vidulich, M.A., and Tsang, P.S. Assessing subjective workload assessment: A comparison of SWAT and the NASA-bipolar methods. Proceedings of the Human Factors Society 29th Annual Meeting, 71–75, 1985.

Vidulich, M.A., and Tsang, P.S. Absolute magnitude estimation and relative judgment approaches to subjective workload assessment. Proceedings of the Human Factors Society 31st Annual Meeting, 1057–1061, 1987.

Vu, K.P.L., Minakata, K., Nguyen, J., Kraut, J., Raza, H., Battiste, V., and Strybel, T.Z. Situation awareness and performance of student versus experienced Air Traffic Controllers. In M.J. Smith and G. Salvendy (Eds.) *Human Interface* (pp. 865–874). Berlin: Springer-Verlag, 2009.

Warvel, L., and Scerbo, M.W. Measurement of mental workload changes during laparoscopy with a visual-spatial task. Proceedings of the Human Factors and Ergonomics Society 59th Annual Meeting, 503–507, 2015.

Weber, F., Haering, C., and Thomaschke, R. Improving the human-computer dialog with increased temporal predictability. *Human Factors* 55(5): 881–892, 2013.

Willems, B., and Heiney, M. *Decision Support Automation Research in the En Route Air Traffic Control Environment (DOT/FAA/CT-TN01/10)*. Atlantic City International Airport, NJ: Federal Aviation Administration William J. Hughes Technical Center, January 2002.

Windell, D., Wiebe, E., Converse-Lane, S., and Beith, B. A comparison of two mental workload instruments in multimedia instruction. Proceedings of the Human Factors and Ergonomics Society 50th Annual Meeting, 1764–1768, 2006.

Wright, J.L., Chen, J.Y.C., Barnes, M.J., and Hancock, P.A. Agent reasoning transparency's effect on operator workload. Proceedings of the Human Factors and Ergonomics Society 60th Annual Meeting, 249–253, 2016.

Yu, D., Abdelrahman, A.M., Buckarma, E.N.H., Lowndes, B.R., Gas, B.L., Finnesgard, E.J., Abdelsattar, J.M., Pandian, T.K., Khatib, M.E., Farley, D.R., and Halbeck, S. Mental and physical workloads in a competitive laparoscopic skills training environment: A pilot study. Proceedings of the Human Factors and Ergonomics Society 59th Annual Meeting, 508–512, 2015.

2.3.3.12 Profile of Mood States

General description – The shortened version of the Profile of Mood States (POMS) scale (Shachem, 1983) provides measures of self-rated tension, depression, anger, vigor, fatigue, and confusion. It has been used as a measure of workload.

Strengths and limitations – Reliability and validation testing of the POMS has been extensive. For example, McNair and Lorr (1964) reported test/retest reliabilities of 0.61 to 0.69 for the six factors. Reviews of the sensitivity and reliability of the POMS have been favorable (Norcross et al., 1984). Constantini et al.(1971) reported significant positive correlations between POMS and the Psychological Screening Inventory, thus yielding consensual validation. Pollock et al. (1979) correlated POMS scales and physiological measures from eight healthy males. The tension and depression scores were significantly correlated with heart rate (+0.75 and +0.76, respectively) and diastolic blood pressure (+0.71 and +0.72, respectively). Heart rate was also significantly correlated with the anger score (+0.70).

The POMS has been used extensively in psychotherapy research (e.g., Haskell et al., 1969; Lorr et al., 1961; McNair et al., 1965; Pugatch et al., 1969) and drug research (e.g., Mirin et al., 1971; Nathan et al., 1970a, 1970b; Pillard and Fisher, 1970).

Storm and Parke (1987) used the POMS to assess the mood effects of a sleep-inducing drug (temazepam) for EF-111 aircrews. As anticipated, there were no significant drug effects on any of the six subscales. Gawron et al. (1988) asked participants to complete the POMS after a 1.75-hour flight. There were no significant crew position effects on rated vigor or fatigue. There was a significant order effect on fatigue, however. Participants who had been pilots first had higher ratings (2.7) than participants who had been copilots first (1.3).

Harris et al. (1995) did not find a significant difference in the fatigue rating between a manual and an automatic tracking group.

Data requirements – The POMS takes about 10 minutes to complete and requires a stiff writing surface. The POMS is available from the Educational and Industrial Testing Service, San Diego, California.

Thresholds – Not stated.

Sources

Costantini, A.F., Braun, J.R., Davis, J.E., and Iervolino, A. The life change inventory: A device for quantifying psychological magnitude of changes experienced by college students. *Psychological Reports* 34(3, Pt. 1): 991–1000, 1971.

Gawron, V.J., Schiflett, S., Miller, J., Ball, J., Slater, T., Parker, F., Lloyd, M., Travale, D., and Spicuzza, R.J. *The Effect of Pyridostigmine Bromide on In-Flight Aircrew Performance (USAFSAM-TR-87-24)*. Brooks Air Force Base, TX: School of Aerospace Medicine, January 1988.

Harris, W.C., Hancock, P.A., Arthur, E.J., and Caird, J.K. Performance, workload, and fatigue changes associated with automation. *International Journal of Aviation Psychology* 5(2): 169–185, 1995.

Haskell, D.H., Pugatch, D., and McNair, D.M. Time-limited psychotherapy for whom? *Archives of General Psychiatry* 21: 546–552, 1969.

Lorr, M., McNair, D.M., Weinstein, G.J., Michaux, W.W., and Raskin, A. Meprobromate and chlorpromazine in psychotherapy. *Archives of General Psychiatry* 4: 381–389, 1961.

McNair, D.M., Goldstein, A.P., Lorr, M., Cibelli, L.A., and Roth, I. Some effects of chlordiazepoxide and meprobromate with psychiatric outpatients. *Psychopharmacologia* 7: 256–265, 1965.

McNair, D.M., and Lorr, M. An analysis of mood in neurotics. *Journal of Abnormal Psychology* 69: 620–627, 1964.

Mirin, S.M., Shapiro, L.M., Meyer, R.E., Pillard, R.C., and Fisher, S. Casual versus heavy use of marijuana: A redefinition of the marijuana problem. *American Journal of Psychiatry* 172: 1134–1140, 1971.

Nathan, P.F., Titler, N.A., Lowenstein, L.M., Solomon, P., and Rossi, A.M. Behavioral analyses of chronic alcoholism: Interaction of alcohol and human contact. *Archives of General Psychiatry* 22: 419–430, 1970a.

Nathan, P.F., Zare, N.C., Ferneau, E.W., and Lowenstein, L.M. Effects of congener differences in alcohol beverages on the behavior of alcoholics. *Quarterly Journal on Studies of Alcohol.* Supplement Number 5: 87–100, 1970b.

Norcross, J.C., Guadagnoli, E., and Prochaska, J.O. Factor structure of the profile of mood states (POMS): Two partial replications. *Journal of Clinical Psychology* 40: 1270–1277, 1984.

Pillard, R.C., and Fisher, S. Aspects of anxiety in dental clinic patients. *Journal of the American Dental Association* 80: 1331–1334, 1970.

Pollock, V., Cho, D.W., Reker, D., and Volavka, J. Profile of mood states: The factors and their correlates. *Journal of Nervous Mental Disorders* 167: 612–614, 1979.

Pugatch, D., Haskell, D.H., and McNair, D.M. Predictors and patterns of change associated with the course of time limited psychotherapy (Mimeo Report), 1969.

Shachem, A. A shortened version of the profile of mood states. *Journal of Personality Assessment*, 47: 305–306, 1983.

Storm, W.F., and Parke, R.C. FB-111A aircrew use of temazepam during surge operations. Proceedings of NATO Advisory Group for Aerospace Research and Development (AGARD) Biochemical Enhancement of Performance Conference (Paper No. 415, pp. 12-1–12-12). Neuilly-sur-Seine, France: AGARD, 1987.

2.3.3.13 Subjective Workload Assessment Technique

General description – The Subjective Workload Assessment Technique (SWAT) combines ratings of three different scales (see Table 2.24) to produce an interval scale of mental workload. These scales are: (1) time load, which reflects the amount of spare time available in planning, executing, and monitoring a task, (2) mental effort load, which assesses how much conscious mental effort and planning are required to perform a task, and (3) psychological stress load, which measures the amounts of risk, confusion, frustration, and anxiety associated with task performance. A more complete description is given in Reid and Nygren (1988). A description of the initial conjoint measurement model for SWAT is described in Nygren (1982, 1983).

Strengths and limitations – SWAT has been found to be a valid (Albery et al., 1987; Haworth et al., 1986; Masline, 1986; Reid et al., 1981a, 1981b; Vidulich

TABLE 2.24
SWAT Scales

Time Load
1. Often have spare time. Interruptions or overlap among activities occur infrequently or not at all.
2. Occasionally have spare time. Interruptions or overlap among activities occur frequently.
3. Almost never have spare time. Interruptions or overlap among activities are frequent or occur all the time.

Mental Effort Load
1. Very little conscious mental effort or concentration required. Activity is almost automatic, requiring little or no attention.
2. Moderate conscious mental effort or concentration required. Complexity of activity is moderately high due to uncertainty, unpredictability, or unfamiliarity. Considerable attention required.
3. Extensive mental effort and concentration are necessary. Very complex activity requiring total attention.

Psychological Stress Load
1. Little confusion, risk, frustration, or anxiety exists and can be easily accommodated.
2. Moderate stress due to confusion, frustration, or anxiety noticeably adds to workload. Significant compensation is required to maintain adequate performance.
3. High to very intense stress due to confusion, frustration, or anxiety. High to extreme determination and self-control required (Potter and Bressler, 1989, pp. 12–14).

and Tsang, 1985, 1987; Warr et al., 1986), sensitive (Eggemeier et al., 1982), reliable (Corwin et al., 1989; Gidcomb, 1985), and relatively unobtrusive (Crabtree et al., 1984; Courtright and Kuperman, 1984; Eggemeier, 1988) measure of workload. Further, SWAT ratings are not affected by delays of up to 30 minutes (Eggemeier et al., 1983) nor by intervening tasks of all but difficult tasks (Eggemeier et al., 1984; Lutmer and Eggemeier, 1990). Moroney et al. (1995) concur. Also, Eggleston (1984) found a significant correlation between projected SWAT ratings made during system concept evaluation and those made during ground-based simulation of the same system.

Warr (1986) reported that SWAT ratings were less variable than Modified Cooper-Harper Rating Scale ratings. Kilmer et al. (1988) reported that SWAT was more sensitive to changes in difficulty of a tracking task than the Modified Cooper-Harper Rating Scale was. Nygren (1991) stated that SWAT provides a good cognitive model of workload, sensitive to individual differences. Anthony and Biers (1997), however, found no difference between Overall Workload Scale and SWAT ratings. Their participants were 48 introductory psychology students performing a memory recall task.

SWAT has been used in diverse environments, for example, test aircraft (Papa and Stoliker, 1988), a high-G centrifuge (Albery et al., 1985; Albery, 1989), command, control, and communications centers (Crabtree et al., 1984), nuclear power plants (Beare and Dorris, 1984), domed flight simulators (Reid et al., 1982; Skelly and Simons, 1983), tank simulators (Whitaker et al., 1989); and the benign laboratory setting (Graham and Cook, 1984; Kilmer et al., 1988).

In the laboratory, SWAT has been used to assess the workload associated with critical tracking and communication tasks (Reid et al., 1981a), memory tasks (Eggemeier et al., 1982; Eggemeier and Stadler, 1984; Potter and Acton, 1985), and monitoring tasks (Notestine, 1984). Hancock and Caird (1993) reported significant increases in SWAT ratings as the shrink rate of the target decreased and as the number of steps from the cursor to the target increased. In a visual choice reaction time task, Cassenti et al. (2011) reported linear increases in SWAT scores as a function of number of concurrent tasks and with decreases in stimulus presentation time.

Usage in simulated flight has also been extensive (Haworth et al., 1986; Nataupsky and Abbott, 1987; Schick and Hann, 1987; Skelly and Purvis, 1985; Skelly et al., 1983; Thiessen et al., 1986; Ward and Hassoun, 1990). Use in military flight simulators has been extensive. For example, Bateman and Thompson (1986) reported that SWAT ratings increased as task difficulty increased. Their data were collected in an aircraft simulator during a tactical mission. Vickroy (1988), also using an aircraft simulator, reported that SWAT ratings increased as the amount of air turbulence increased. Fracker and Davis (1990) reported significant increases in SWAT as the number of simulated enemy aircraft increased from 1 to 3. Hankey and Dingus (1990) reported that SWAT was sensitive to changes in time on task and fatigue. Hancock et al. (1995) reported that SWAT was highly correlated with the difficulty of a simulated flight task. However, See and Vidulich (1997) reported significant effects of target type and threat status on SWAT scores in a combat aircraft simulator. There were no significant correlations of SWAT with overall workload but two subscales correlated with peak workload (effort, $r = +0.78$; stress, $r = +0.76$). Vidulich (1991) reported test-retest reliability of +0.606 in SWAT ratings for tracking, choice RT, and Sternberg tasks.

Arbak et al. (1984) applied SWAT in a reflective manner based on mission performance of B-52 pilots and copilots. The authors concluded this reflective manner was useful when a two-on-one interview technique was applied, the original situation was described in detail, segment boundaries are well identified, and instances reviewed impact performance. Kuperman and Wilson (1985) applied SWAT projectively early in system design.

Usage in actual flight has also been extensive. For example, Pollack (1985) used SWAT to assess differences in workload between flight segments. She reported that C-130 pilots had the highest SWAT scores during the approach segment of the mission. She also reported higher SWAT ratings during the preflight segments of tactical, rather than proficiency, missions. Haskell and Reid (1987) found significant differences in SWAT ratings between flight maneuvers and also between successfully completed maneuvers and those that were not successfully completed. Gawron et al. (1988) analyzed SWAT ratings made by the pilot and copilot four times during each familiarization and data flight: (1) during the taxi out to the runway, (2) just prior to a simulated air drop, (3) just after a simulated air drop, and (4) during the taxi back to the hangar. There were significant segments effects. Specifically, SWAT ratings were highest before the drop and lowest for preflight. The ratings during postdrop and postflight were both moderate.

Experience with SWAT has not been all positive, however. For example, Boyd (1983) reported that there were significant positive correlations between the three workload scales in a text-editing task. This suggests that the three dimensions of workload are not independent. This, in turn, poses a problem for use of conjoint measurement techniques. Further, Derrick (1983) and Hart (1986) suggested that three scales may not be adequate for assessing workload. In examining the three scales, Biers and Masline (1987) compared three alternative analysis methods for SWAT: conjoint analysis, simple sum of the three subscales, and a weighted linear combination. They reported that the individual scales were differentially sensitive to different task demands. Masline and Biers (1987) also reported greater correspondence between projective and post-task workload ratings using magnitude estimation than either SWAT or equal-appearing intervals. Further, Battiste and Bortolussi (1988) reported a test/retest correlation of +0.751 but also stated that, of the 144 SWAT ratings reported during a simulated B-727 flight, 59 were zero. In commercial aircraft operations, Corwin (1989) reported no difference between in-flight and post-flight ratings of SWAT in only two of three flight conditions.

There have been other inconsistent results. For example, Haworth et al. (1986) reported that, although the SWAT was able to discriminate control configuration conditions in a single-pilot configuration, it could not discriminate these same conditions in a pilot/copilot configuration. Wilson et al. (1990) reported no significant differences in SWAT ratings among display formats, in contrast to pilot comments. van de Graaff (1987) reported considerable (60 points) intersubject variability in SWAT ratings during an in-flight approach task. Hill et al. (1992) reported that SWAT was not as sensitive to workload or as accepted by Army operators as NASA TLX and the Overall Workload Scale.

Vidulich and Tsang (1986) reported that SWAT failed to detect resource competition effects in a dual-task performance of tracking and a directional transformation task. Rueb et al. (1992) reported that only one of three difficult simulated aerial refueling missions had SWAT scores above the 40 redline.

Vidulich and Pandit (1987) concluded that SWAT was not an effective measure of individual differences. This conclusion was based on no significant correlations of SWAT with any of the scales on the Jenkins Activity Survey, Rotter's Locus of Control, the Cognitive Failures Questionnaire, the Cognitive Interference Questionnaire, the Thought Occurrence Questionnaire, the California Q-Sort, and the Myers-Briggs Type Indicator.

Reid (1985) warned that SWAT was most sensitive when workload was moderate to high. Acton and Rokicki (1986) surveyed SWAT users in the Air Force Test and Evaluation community and suggested the development of a user's guide to help train raters. Additionally, they suggested the development of guidelines for task selection and methods to handle small data sets. Further, Nygren et al. (1998) reported that how individuals weighted the SWAT dimensions affected their workload ratings. Their participants were 124 introductory psychology students who were categorized into one of six groups based on their SWAT dimension weightings.

Svensson et al. (1997) reported the reliability of SWAT to be +0.74 among 18 pilots flying simulated low-level, high-speed missions. The correlation with the Bedford Workload Scale was +0.687 and with the NASA TLX was +0.735. Rubio et al. (2004) reported that SWAT was less sensitive to single versus dual task workload than the Workload Profile (WP). Their participants were 36 psychology students who performed Sternberg's Memory Searching Task and/or a tracking task. There was no significant effect of memory set size or path width on the tracking task for SWAT in the single-task conditions. In the dual task, there was only a significant effect of path width.

A Simplified SWAT (S-SWAT) was used by Morgan and Hancock (2010) to measure driver workload in a simulator. S-SWAT has three scales: time, mental effort, and psychological stress each with a range of 0 to 100. An unweighted average of these three scores was used as a measure of mental workload. The authors reported a significant increase then slight decrease in workload over the drive. Workload was highest when a navigation system failed. In looking at the scores separately, there was a significant difference in time demand reflecting the same effect as the average of all three scores. Mental effort was also significantly higher in the middle of the drive but did not decrease at the end. Similar results were reported for psychological stress.

Thresholds – The minimum value is 0 and the maximum value is 100. High workload is associated with the maximum value. In addition, ratings of the time, effort, and stress scales may be individually examined as workload components (Eggemeier et al., 1983). Colle and Reid (2005) reported a redline value of 41.1, which was within the 40 +/− 10 suggested by Reid and Colle (1988).

Data requirements – SWAT requires two steps to use: scale development and event scoring. Scale development requires participants to rank, from lowest to highest workload, 27 combinations of three levels of the three workload subscales. The levels of each subscale are presented in Table 29. Reid et al. (1982) describe their individual differences approach to scale development. Programs to calculate the SWAT score for every combination of ratings on the three subscales are available from the Air Force Research Laboratory at Wright-Patterson Air Force Base. A user's manual is also available from the same source.

During event scoring, the participant is asked to provide a rating (1, 2, 3) for each subscale. The experimenter then maps the set of ratings to the SWAT score (1 to 100) calculated during the scale development step. Haskell and Reid (1987) suggests that the tasks to be rated be meaningful to the participants and, further, that the ratings not interfere with performance of the task. Acton and Colle (1984) reported that the order in which the subscale ratings are presented does not affect the SWAT score. However, it is suggested that the order remain constant to minimize confusion. Eggleston and Quinn (1984) recommended developing a detailed system and operating environment description for prospective ratings. Finally, Biers and McInerney (1988) reported that the card sort did not affect the task workload ratings and therefore may not be necessary.

Luximon and Goonetilleke (2001) compared the traditional SWAT with five variants. The variants used the continuous SWAT subscales but a pairwise comparison of the subscales. The variants were discrete SWAT dimensions, continuous SWAT dimensions with the minimum weight equal to zero, continuous SWAT dimensions with non-zero minimum weight, continuous SWAT dimensions with equal weight, and continuous SWAT dimensions with weight based on principal component analysis. Based on the data from 15 participants performing math at varying levels of difficulty, the authors concluded that the traditional SWAT was the least sensitive to workload and equal weight and the principal component variants the most sensitive.

Gidcomb (1985) reported casual card sorts and urged emphasizing the importance of the card sort to SWAT raters. A computerized version of the traditional card sort was developed at the Air Force School of Aerospace Medicine. This version eliminates the tedium and dramatically reduces the time to complete the SWAT card sort.

Sources

Acton, W., and Colle, H. The effect of task type and stimulus pacing rate on subjective mental workload ratings. Proceedings of the IEEE 1984 National Aerospace and Electronics Conference, 818–823, 1984.

Acton, W.H., and Rokicki, S.M. Survey of SWAT use in operational test and evaluation. Proceedings of the Human Factors Society 30th Annual Meeting, 1221–1224, 1986.

Albery, W. B. The effect of sustained acceleration and noise on workload in human operators. *Aviation, Space, and Environmental Medicine* 6(10): 943–948, 1989.

Albery, W., Repperger, D., Reid, G., Goodyear, C., and Roe, M. Effect of noise on a dual task: Subjective and objective workload correlates. Proceedings of the National Aerospace and Electronics Conference, 1457–1463, 1987.

Albery, W.B., Ward, S.L., and Gill, R.T. Effect of acceleration stress on human workload (Technical Report AMRL-TR-85-039). Wright-Patterson Air Force Base, OH: Aerospace Medical Research Laboratory, May 1985.

Anthony, C.R., and Biers, D.W. Unidimensional versus multidimensional workload scales and the effect of number of rating scale categories. Proceedings of the Human Factors and Ergonomics Society 41st Annual Meeting, 1084–1088, 1997.

Arbak, C.J., Shew, R.L., and Simons, J.C. The use of reflective SWAT for workload assessment. Proceedings of the Human Factors Society 28th Annual Meeting, 959–962, 1984.

Bateman, R.P., and Thompson, M.W. Correlation of predicted workload with actual workload using the subjective workload assessment technique. Proceedings of the SAE AeroTech Conference, 1986.

Battiste, V., and Bortolussi, M.R. Transport pilot workload: A comparison of two subjective techniques. Proceedings of the Human Factors Society 32nd Annual Meeting, 150–154, 1988.

Beare, A., and Dorris, R. The effects of supervisor experience and the presence of a shift technical advisor on the performance of two-man crews in a nuclear power plant simulator. Proceedings of the Human Factors Society 28th Annual Meeting, 242–246, 1984.

Biers, D.W., and Masline, P.J. Alternative approaches to analyzing SWAT data. Proceedings of the Human Factors Society 31st Annual Meeting, 63–66, 1987.

Biers, D.W., and McInerney, P. An alternative to measuring subjective workload: Use of SWAT without the card sort. Proceedings of the Human Factors Society 32nd Annual Meeting, 1136–1139, 1988.

Boyd, S.P. Assessing the validity of SWAT as a workload measurement instrument. Proceedings of the Human Factors Society 27th Annual Meeting, 124–128, 1983.

Cassenti, D.N., Kelley, T.D., Colle, H.A., and McGregor, E.A. Modeling performance measures and self-ratings of workload in a visual scanning task. Proceedings of the Human Factors and Ergonomics Society 55th Annual Meeting, 870–874, 2011.

Colle, H. A., and Reid, G.B. Estimating a mental workload redline in a simulated air-to-ground combat mission. *The International Journal of Aviation Psychology* 15(4): 303–319, 2005.

Corwin, W.H. In-flight and post-flight assessment of pilot workload in commercial transport aircraft using SWAT. Proceedings of the 5th Symposium on Aviation Psychology, 808–813, 1989.

Corwin, W.H., Sandry-Garza, D.L., Biferno, M.H., Boucek, G.P., Logan, A.L., Jonsson, J.E., and Metalis, S.A. *Assessment of Crew Workload Measurement Methods, Techniques, and Procedures. Volume I – Process Methods and Results (WRDC-TR-89-7006).* Wright-Patterson Air Force Base, OH, September 1989.

Courtright J.F., and Kuperman, G. Use of SWAT in USAF system T&E. Proceedings of the Human Factors Society 28th Annual Meeting, 700–703, 1984.

Crabtree, M.A., Bateman, R.P., and Acton, W. Benefits of using objective and subjective workload measures. Proceedings of the Human Factors Society 28th Annual Meeting, 950–953, 1984.

Derrick, W.L. Examination of workload measures with subjective task clusters. Proceedings of the Human Factors Society 27th Annual Meeting, 134–138, 1983.

Eggemeier, F.T. Properties of workload assessment techniques. In P.A. Hancock and N. Meshtaki (Eds.) *Human Mental Workload* (pp. 41–62). Amsterdam: North-Holland, 1988.

Eggemeier, F.T., Crabtree, M.S., and LaPointe, P. The effect of delayed report on subjective ratings of mental workload. Proceedings of the Human Factors Society 27th Annual Meeting, 139–143, 1983.

Eggemeier, F.T., Crabtree, M.S., Zingg, J.J., Reid, G.B., and Shingledecker, C.A. Subjective workload assessment in a memory update task. Proceedings of the Human Factors Society 26th Annual Meeting, 643–647, 1982.

Eggemeier, F.T., McGhee, J.Z., and Reid, G.B. The effects of variations in task loading on subjective workload scales. Proceedings of the IEEE 1983 National Aerospace and Electronics Conference, 1099–1106, 1983.

Eggemeier, F.T., Melville, B., and Crabtree, M. The effect of intervening task performance on subjective workload ratings. Proceedings of the Human Factors Society 28th Annual Meetings, 954–958, 1984.

Eggemeier, F.T., and Stadler, M. Subjective workload assessment in a spatial memory task. Proceedings of the Human Factors Society 28th Annual Meeting, 680–684, 1984.

Eggleston, R.G. A comparison of projected and measured workload ratings using the subjective workload assessment technique (SWAT). Proceedings of the National Aerospace and Electronics Conference, 827–831, 1984.

Eggleston, R.G., and Quinn, T.J. A preliminary evaluation of a projective workload assessment procedure. Proceedings of the Human Factors Society 28th Annual Meeting, 695–699, 1984.

Fracker, M.L., and Davis, S.A. Measuring operator situation awareness and mental workload. Proceedings of the 5th Mid-Central Ergonomics/Human Factors Conference, 23–25, 1990.

Gawron, V.J., Schiflett, S., Miller, J., Ball, J., Slater, T., Parker, F., Lloyd, M., Travale, D., and Spicuzza, R.J. *The Effect of Pyridostigmine Bromide on In-Flight Aircrew Performance (USAFSAM-TR-87-24).* Brooks Air Force Base, TX: School of Aerospace Medicine, January 1988.

Gidcomb, C. *Survey of SWAT Use in Flight Test (BDM/A-85-0630-7R).* Albuquerque, NM: BDM Corporation, 1985.

Graham, C.H., and Cook, M.R. *Effects of Pyridostigmine on Psychomotor and Visual Performance (AFAMRL-TR-84-052).* Wright-Patterson Air Force Base, OH: Armstrong Aerospace Medical Research Laboratory, September 1984.

Hancock, P.A., and Caird, J.K. Experimental evaluation of a model of mental workload. *Human Factors* 35(3): 413–419, 1993.

Hancock, P.A., Williams, G., Manning, C.M., and Miyake, S. Influence of task demand characteristics on workload and performance. *International Journal of Aviation Psychology* 5(1): 63–86, 1995.

Hankey, J.M., and Dingus, T.A. A validation of SWAT as a measure of workload induced by changes in operator capacity. Proceedings of the Human Factors Society 34th Annual Meeting, 113–115, 1990.

Hart, S.G. Theory and measurement of human workload. In J. Seidner (Ed.) *Human Productivity Enhancement* (vol. 1, pp. 396–455). New York: Praeger, 1986.

Haskell, B.E., and Reid, G.B. The subjective perception of workload in low-time private pilots: A preliminary study. *Aviation, Space, and Environmental Medicine* 58: 1230–1232, 1987.

Haworth, L.A., Bivens, C.C., and Shively, R.J. An investigation of single-piloted advanced cockpit and control configuration for nap-of-the-earth helicopter mission tasks. Proceedings of the 42nd Annual Forum of the American Helicopter Society, 657–671, 1986.

Hill, S.G., Iavecchia, H.P., Byers, J.C., Bittner, A.C., Zaklad, A.L., and Christ, R.E. Comparison of four subjective workload rating scales. *Human Factors* 34: 429–439, 1992.

Kilmer, K.J., Knapp, R., Burdsal, C., Borresen, R., Bateman, R.P., and Malzahn, D. A comparison of the SWAT and modified Cooper-Harper scales. Proceedings of the Human Factors 32nd Annual Meeting, 155–159, 1988.

Kuperman, G.G., and Wilson, D.L. A workload analysis for strategic conventional standoff capability missions. Proceedings of the Human Factors Society 29th Annual Meeting, 635–639, 1985.

Lutmer, P.A., and Eggemeier, F.T. The effect of intervening task performance and multiple ratings on subjective ratings of mental workload. Paper presented at the 5th Mid-Central Ergonomics Conference, University of Dayton, Dayton, OH, 1990.

Luximon, A., and Goonetilleke, R.S. Simplified subjective workload assessment technique. *Ergonomics* 44(3): 229–243, 2001.

Masline, P.J. A comparison of the sensitivity of interval scale psychometric techniques in the assessment of subjective mental workload. Unpublished master's thesis, University of Dayton, Dayton, OH, 1986.

Masline, P.J., and Biers, D.W. An examination of projective versus post-task subjective workload ratings for three psychometric scaling techniques. Proceedings of the Human Factors Society 31st Annual Meeting, 77–80, 1987.

Morgan, J.F., and Hancock, P.A. The effect of prior task loading on mental workload: An example of hysteresis in driving. *Human Factors* 53(1): 75–86, 2010.

Moroney, W.F., Biers, D.W., and Eggemeier, F.T. Some measurement and methodological considerations in the application of subjective workload measurement techniques. *International Journal of Aviation Psychology* 5(1): 87–106, 1995.

Nataupsky, M., and Abbott, T.S. Comparison of workload measures on computer-generated primary flight displays. Proceedings of the Human Factors Society 31st Annual Meeting, 548–552, 1987.

Notestine, J. Subjective workload assessment and effect of delayed ratings in a probability monitoring task. Proceedings of the Human Factors Society 28th Annual Meeting, 685–690, 1984.

Nygren, T.E. *Conjoint Measurement and Conjoint Scaling: A User's Guide (AFAMRL-TR-82–22)*. Wright-Patterson Air Force Base, OH: Aerospace Medical Research Laboratory, April 1982.

Nygren, T.E. *Investigation of an Error Theory for Conjoint Measurement Methodology (763025/714404)*. Columbus, OH: Ohio State University Research Foundation, May 1983.

Nygren, T.E. Psychometric properties of subjective workload measurement techniques: Implications for their use in the assessment of perceived mental workload. *Human Factors* 33: 17–33, 1991.

Nygren, T.E., Schnipke, S., and Reid, G. Individual differences in perceived importance of SWAT workload dimensions: Effects on judgment and performance in a virtual high workload environment. Proceedings of the Human Factors and Ergonomics Society 42nd Annual Meeting, 816–820, 1998.

Papa, R.M., and Stoliker, J.R. *Pilot Workload Assessment: A Flight Test Approach*. Washington, DC: American Institute of Aeronautics and Astronautics, 88-2105, 1988.

Pollack, J. *Project Report: An Investigation of Air Force Reserve Pilots' Workload*. Dayton, OH: Systems Research Laboratory, November 1985.

Potter, S.S., and Acton, W. Relative contributions of SWAT dimensions to overall subjective workload ratings. Proceedings of 3rd Symposium on Aviation Psychology, 231–238, 1985.

Potter, S.S., and Bressler, J.R. *Subjective Workload Assessment Technique (SWAT): A User's Guide*. Wright-Patterson Air Force Base, OH: Armstrong Aerospace Medical Research Laboratory, July 1989.

Reid, G.B. Current status of the development of the Subjective Workload Assessment Technique. Proceedings of the Human Factors Society 29th Annual Meeting, 220–223, 1985.

Reid, G.B., and Colle, H.A. Critical SWAT values for predicting operator workload. Proceedings of the Human Factors Society 32nd Annual Meeting, 1414–1418, 1988.

Reid, G.B., Eggemeier, F., and Nygren, T. An individual differences approach to SWAT scale development. Proceedings of the Human Factors Society 26th Annual Meeting, 639–642, 1982.

Reid, G.B., Eggemeier, F.T., and Shingledecker, C.A. In M.L. Frazier and R.B. Crombie (Eds.) Proceedings of the Workshop on Flight Testing to Identify Pilot Workload and Pilot Dynamics (AFFTC-TR-82-5) (pp. 281–288). Edwards AFB, CA, May 1982.

Reid, G.B., and Nygren, T.E. The subjective workload assessment technique: A scaling procedure for measuring mental workload. In P.A. Hancock and N. Mehtaki (Eds.) *Human Mental Workload* (pp. 185–218). Amsterdam: North Holland, 1988.

Reid, G.B., Shingledecker, C.A., and Eggemeier, F.T. Application of conjoint measurement to workload scale development. Proceedings of the Human Factors Society 25th Annual Meeting, 522–526, 1981a.

Reid, G.B., Shingledecker, C.A., Nygren, T.E., and Eggemeier, F.T. Development of multidimensional subjective measures of workload. Proceedings of the IEEE International Conference on Cybernetics and Society, 403–406, 1981b.

Rubio, S., Diaz, E., Martin, J., and Puente, J.M. Evaluation of subjective mental workload: A comparison of SWAT, NASA-TLX, and Workload Profile Methods. *Applied Psychology: An International Review* 53(1): 61–86, 2004.

Rueb, J., Vidulich, M., and Hassoun, J.A. Establishing workload acceptability: An evaluation of a proposed KC-135 cockpit redesign. Proceedings of the Human Factors Society 36th Annual Meeting, 17–21, 1992.

Schick, F.V., and Hann, R.L. The use of subjective workload assessment technique in a complex flight task. In A.H. Roscoe (Ed.) *The Practical Assessment of Pilot Workload*, AGARDograph No. 282 (pp. 37–41). Neuilly-sur-Seine, France: AGARD, 1987.

See, J.E., and Vidulich, M.A. Assessment of computer modeling of operator mental workload during target acquisition. Proceedings of the Human Factors and Ergonomics Society 41st Annual Meeting, 1303–1307, 1997.

Skelly, J.J., and Purvis, B. B-52 wartime mission simulation: Scientific precision in workload assessment. Paper presented at the 1985 Air Force Conference on Technology in Training and Education, Colorado Springs, CO, April 1985.

Skelly, J.J., Reid, G.B., and Wilson, G.R. B-52 full mission simulation: Subjective and physiological workload applications. Paper presented at the Second Aerospace Behavioral Engineering Technology Conference, 1983.

Skelly, J.J., and Simons, J.C. Selecting performance and workload measures for full-mission simulation. Proceedings of the IEEE 198 National Aerospace and Electronics Conference, 1082–1085, 1983.

Svensson, E., Angelborg-Thanderz, M., Sjoberg, L., and Olsson, S. Information complexity – Mental workload and performance in combat aircraft. *Ergonomics* 40(3): 362–380, 1997.

Thiessen, M.S., Lay, J.E., and Stern, J.A. *Neuropsychological Workload Test Battery Validation Study (FZM 7446)*. Fort Worth, TX: General Dynamics, 1986.

van de Graaff, R.C. An in-flight investigation of workload assessment techniques for civil aircraft operations (NLR-TR-87119 U). Amsterdam, the Netherlands: National Aerospace Laboratory, 1987.

Vickroy, S.C. *Workload Prediction Validation Study: The Verification of CRAWL Predictions*. Wichita, KS: Boeing Military Airplane Company, 1988.

Vidulich, M.A. The Bedford scale: Does it measure spare capacity? Proceedings of the 6th International Symposium on Aviation Psychology, 1136–1141, 1991.

Vidulich, M.A., and Pandit, P. Individual differences and subjective workload assessment: Comparing pilots to nonpilots. Proceedings of the 4th International Symposium on Aviation Psychology, 630–636, 1987.

Vidulich, M.A., and Tsang, P.S. Techniques of subjective workload assessment: A comparison of two methodologies. Proceedings of the 3rd International Symposium on Aviation Psychology, 239–246, 1985.

Vidulich, M.A., and Tsang, P.S. Techniques of subjective workload assessment: A comparison of SWAT and NASA-Bipolar methods. *Ergonomics* 29(11): 1385–1398, 1986.

Vidulich, M.A., and Tsang, P.S. Absolute magnitude estimation and relative judgment approaches to subjective workload assessment. Proceedings of the Human Factors Society 31st Annual Meeting, 1057–1061, 1987.

Ward, G.F., and Hassoun, J.A. *The Effects of Head-Up Display (HUD) Pitch Ladder Articulation, Pitch Number Location and Horizon Line Length on Unusual Altitude Recoveries for the F-16 (ASD-TR-90-5008)*. Wright-Patterson Air Force Base, OH: Crew Station Evaluation Facility, July 1990.

Warr, D.T. A comparative evaluation of two subjective workload university measures: The subjective assessment technique and the Modified Cooper-Harper rating. Master's thesis. Dayton, OH: Wright State, 1986.

Warr, D., Colle, H., and Reid, G.B. A comparative evaluation of two subjective workload measures: The subjective workload assessment technique and the Modified Cooper-Harper Scale. Paper presented at the Symposium on Psychology in the Department of Defense. US Air Force Academy, Colorado Springs, CO, 1986.

Whitaker, L.A., Peters, L., and Garinther, G. Tank crew performance: Effects of speech intelligibility on target acquisition and subjective workload assessment. Proceedings of the Human Factors Society 33rd Annual Meeting, 1411–1413, 1989.

Wilson, G.F., Hughes, E., and Hassoun, J. A physiological and subjective evaluation of a new aircraft display. Proceedings of the Human Factors Society 34th Annual Meeting, 1441–1443, 1990.

2.3.3.14 Team Workload Questionnaire

General description – The Team Workload Questionnaire (TWLQ) includes the six scales of the NASA TLX as well as Emotional Demand, Performance Monitor Demand, and Teamwork (Sellers et al., 2014). The scales are presented in Table 2.25. Each scale ranges from 0 (very low) to 10 (very high). Responses are multiplied by 10 so each item varies between 0 and 100.

TABLE 2.25

Team Workload Questionnaire (Sellers et al., 2014, p. 991)

Scale	Definition
Emotion demand	How much did the task require you to control your emotions (e.g., anger, joy, disappointment)?
Performance monitor demand	How much did the task require you to monitor your performance (i.e., ensure you were performing at specific levels?)
Communication demand	How much communication activity was required (e.g., discussing, negotiating, sending and receiving messages, etc.)
Coordination demand	How much coordination activity was required (e.g., correction, adjustment, etc.)?
Time share demand	How difficult was it to share and manage time between task-work (work done individually) and team-work (work done as a team)?
Team effectiveness	How successful do you think the team was in working together?
Team support	How difficult was it to provide and receive support (providing guidance, helping team members, providing instructions, etc.) from team members?
Team dissatisfaction	How irritated and annoyed were you with your team?
Team emotion demand	How emotionally demanding was working in the team?
Team performance monitoring demand	How much did the task require you to monitor your team's performance?

Strengths and limitations – (Sellers et al. (2014), based on responses from 216 members of a sports team, reported 57.80% of the variance accounted for in a three-factor model: task workload ($\alpha = 0.783$), team workload ($\alpha = 0.739$), and team-task balancing ($\alpha = 0.692$). Sellers et al. (2015) asked 14 teams of two to complete the TWLQ after controlling a quadcoptor UAV. They reported significant corrections between task workload and team workload ($r = +0.673$), team workload and task-team balancing ($r = +0.484$), and task-team balancing and performance ($r = +0.465$).

In a more recent study, Greenlee et al. (2017) reported that TWLQ was a poor fit to assess team workload in a personnel-hiring task.

Thresholds – Each item varies between 0 and 100.

Sources

Greenlee, E.T., Funke, G.J., and Rice, L. Evaluation of the Team Workload Questionnaire (TWQ) in a team choice task. Proceedings of the 61st Human Factors and Ergonomics Society Annual Meeting, 1317, 2017.

Sellers, J., Helton, W.S., Naswall, K., Funke, G.J., and Knott, B.A. Development of the Team Workload Questionnaire (TWLQ). Proceedings of the Human Factors and Ergonomics Society 58th Annual Meeting, 989–993, 2014.

Sellers, J., Helton, W.S., Naswall, K., Funke, G.J., and Knott, B.A. The Team Workload Questionnaire (TWLQ): A simulated unmanned vehicle task. Proceedings of the Human Factors and Ergonomics Society 59th Annual Meeting, 1382–1386, 2015.

2.3.3.15 Workload/Compensation/Interference/Technical Effectiveness

General description – The Workload/Compensation/Interference/Technical Effectiveness (WCI/TE) rating scale (see Figure 2.16) requires participants to rank the 16 matrix cells and then rate specific tasks. The ratings are converted by conjoint scaling techniques to values of 0 to 100.

Strengths and limitations – Wierwille and Connor (1983) reported sensitivity of WCI/TE ratings to three levels of task difficulty in a simulated flight task. Wierwille et al. (1985a) reported sensitivity to changes in difficulty in psychomotor, perceptual, and mediational tasks. Wierwille et al. (1985b) reported that WCI/TE was sensitive to variations in the difficulty of a secondary mathematical task during a simulated flight task. However, O'Donnell and Eggemeier (1986) suggest that the WCI/TE should not be used as a direct measure of workload.

Data requirements – Participants must rank the 16 matrix cells and then rate specific tasks. Complex mathematical processing is required to convert the ratings to WCI/TE values.

Thresholds – 0 is minimum workload; 100 is maximum workload.

Human Workload

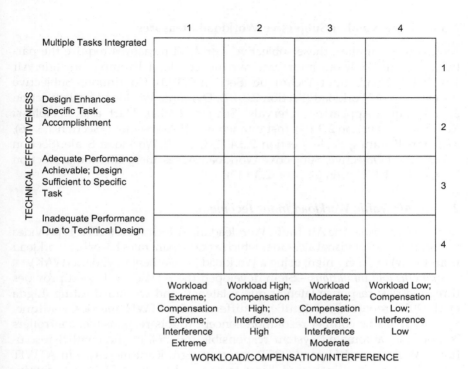

FIGURE 2.16
WCI/TE scale matrix.

Sources

Lysaght, R.J., Hill, S.G., Dick, A.O., Plamondon, B.D., Linton, P.M., Wierwille, W.W., Zaklad, A.L., Bittner, A.C., and Wherry, R.J. Operator workload: Comprehensive review and evaluation of operator workload methodologies (Technical Report 851). Alexandria, VA: Army Research Institute for the Behavioral and Social Sciences, June 1989.

O'Donnell, R.D., and Eggemeier, F.T. Workload assessment methodology. In K.R. Boff, L. Kaufman, and J.P. Thomas (Eds.) *Handbook of Perception and Human Performance. Volume (Vol) 2, Cognitive Processes and Performance.* New York: Wiley, 1986.

Wierwille, W.W., Casali, J.G., Connor, S.A., and Rahimi, M. Evaluation of the sensitivity and intrusion of mental workload estimation techniques. In W. Roner (Ed.) *Advances in Man-Machine Systems Research*, vol. 2 (pp. 51–127). Greenwich, CT: J.A.I. Press, 1985a.

Wierwille, W.W., Rahimi, M., and Casali, J.G. Evaluation of 16 measures of mental workload using a simulated flight task emphasizing mediational activity. *Human Factors* 27(5): 489–502, 1985b.

Wierwille, W.W., and Connor, S.A. Evaluation of twenty workload assessment measures using a psychomotor task in a motion-base aircraft simulation. *Human Factors* 25: 1–16, 1983.

2.3.4 Single Number Subjective Workload Measures

As the name implies, these subjective workload measures require the participant to give only one number to rate the workload. Examples include: Air Traffic Workload Input Technique (Section 2.3.4.1), Continuous Subjective Assessment of Workload (Section 2.3.4.2), Dynamic Workload Scale (Section 2.3.4.3), Equal-Appearing Intervals (Section 2.3.4.4), Hart and Bortolussi Rating Scale (Section 2.3.4.5), Instantaneous Self Assessment (Section 2.3.4.6), McDonnell Rating Scale (Section 2.3.4.7), Overall Workload Scale (Section 2.3.4.8), Pilot Objective/Subjective Workload Assessment Technique (Section 2.3.4.9), and Utilization (Section 2.3.4.10).

2.3.4.1 Air Traffic Workload Input Technique

General description – The Air Traffic Workload Input Technique (ATWIT) provides an auditory and/or visual alert after which a participant rates his or her workload from 1 (low) to 7 (very high) using a Workload Assessment Keyboard (WAK).

Strengths and limitations – Stein developed the ATWIT in 1985. It provides three metrics: response latency, query latency, and workload rating. Ligda et al. (2010) reported no significant difference in ATWIT metrics as a function of air traffic conflict resolution concept (pilots responsible, controllers responsible, automated system responsible for 75% of the conflict resolutions). Willems and Heiney (2002) reported significant increases in ATWIT ratings as task load increased. Time to respond to the ATWIT was significantly faster on the radar side than on the data controller side. The participants were 16 Air Traffic Controllers evaluating a decision support system. Truitt (2013) reported no significant differences in WAK rating for either the ground controller or the local controller for an evaluation of voice, 40% data communications, or 75% data communications.

Loft et al. (2015) compared SPAM, SAGAT, ATWIT, NASA TLX, and SART ratings from 117 undergraduates performing three submarine tasks: contact classification, closest point of approach, and emergency surface. SPAM did not significantly correlate with SART but did with ATWIT and NASA TLX. Fincannon and Ahlstrom (2016), based on a meta-analysis of 15 studies, reported higher correlation of the number of aircraft and the 10-point scale rather than the 7-point scale.

Thresholds – The minimum is 1 (low workload). The maximum is 7 (very high workload).

Sources

Fincannon, T., and Ahlstrom, V. Scale size of the Air Traffic Workload Input Technique (ATWIT): A review of the research. Proceedings of the Human Factors and Ergonomics Society 60th Annual Meeting, 2071–2075, 2016.

Ligda, S.V., Dao, A.V., Vu, K., Strybel, T.Z., Battiste, V., and Johnson, W.W. Impact of conflict avoidance responsibility allocation on pilot workload in a distributed air traffic management system. Proceedings of the Human Factors and Ergonomics Society 54th Annual Meeting, 55–59, 2010.

Loft, S., Bowden, V., Braithwaite, J., Morrell, D.B., Huf, S., and Durso, F.T. Situation awareness measures for simulated submarine track management. *Human Factors* 57(2): 298–310, 2015.

Stein, E.S. *Air Traffic Controller Workload: An Examination of Workload Probe (DOT/FAA/CT-TN84/24).* Atlantic City, NJ: Federal Aviation Administration, April 1985.

Truitt, T.R. An empirical study of digital taxi clearances for departure aircraft. *Air Traffic Control Quarterly* 21(2): 125–151, 2013.

Willems, B., and Heiney, M. *Decision Support Automation Research in the En Route Air Traffic Control Environment (DOT/FAA/CT-TN01/10).* Atlantic City International Airport, NJ: Federal Aviation Administration William J. Hughes technical Center, January 2002.

2.3.4.2 Continuous Subjective Assessment of Workload

General description – The Continuous Subjective Assessment of Workload (C-SAW) requires participants to provide ratings of 1 to 10 (corresponding to the Bedford Workload Scale descriptors) while viewing a videotape of their flight immediately after landing. Computer prompts ratings at rates up to once every three seconds. A bar-chart or graph against the timeline is the output.

Strengths and limitations – Jensen (1995) stated that participants could reliably provide ratings every three seconds. He reported C-SAW was sensitive to differences between a HUD and a head-down display. C-SAW has high face validity but has not been formally validated.

Thresholds – The minimum is zero.

Source

Jensen, S.E. Developing a flight workload profile using Continuous Subjective Assessment of Workload (C-SAW). Proceedings of the 21st Conference of the European Association for Aviation Psychology, Chapter 46, 1995.

2.3.4.3 Dynamic Workload Scale

General description – The Dynamic Workload Scale is a seven-point workload scale (see Figure 2.17) developed as a tool for aircraft certification. It has been used extensively by Airbus Industries.

Strengths and limitations – Speyer et al. (1987) reported high concordance between pilot and observer ratings as well as sensitivity to workload increases.

Workload Assessment		CRITERIA			Appreciation
		Reserve Capacity	Interruptions	Effort or Stress	
Light	2	Ample			Very Acceptable
Moderate	3	Adequate	Some		Well Acceptable
Fair	4	Sufficient	Recurring	Not Undue	Acceptable
High	5	Reduced	Repetitive	Marked	High but Acceptable
Heavy	6	Little	Frequent	Significant	Just Acceptable
Extreme	7	None	Continuous	Acute	Not Acceptable Continuously
Supreme	8	Impairment	Impairment	Impairment	Not Acceptable Instantaneously

FIGURE 2.17
Dynamic workload scale.

Data requirements – Dynamic Workload Scale ratings must be given by both a pilot and an observer-pilot. The pilot is cued to make a rating; the observer gives a rating whenever workload changes or five minutes have passed. Hall et al. (2010) also used a seven-point workload scale recorded every five minutes. They reported significantly lower workload for a prototype command and control display than for the baseline current display.

Thresholds – Two is minimum workload; eight, maximum workload.

Sources

Hall, D.S., Shattuck, L.G., and Bennett, K.B. Evaluation of an ecological interface designed for military command and control. Proceedings of the Human Factors and Ergonomics Society 54th Annual Meeting, 423–427, 2010.

Speyer, J., Fort, A., Fouillot, J., and Bloomberg, R. Assessing pilot workload for minimum crew certification. In A.H. Roscoe (Ed.) *The Practical Assessment of Pilot Workload (AGARDograph Number 282)* (pp. 90–115). Neuilly-sur-Seine, France: AGARD, 1987.

2.3.4.4 Equal-Appearing Intervals

General description – Participants rate the workload in one of several categories using the assumption that each category is equidistant from adjacent categories.

Strengths and limitations – Hicks and Wierwille (1979) reported sensitivity to task difficulty in a driving simulator. Masline (1986) reported comparable

results with the magnitude estimates and SWAT ratings but greater ease of administration. Masline, however, warned of rater bias.
Data requirements – Equal intervals must be clearly defined.
Thresholds – Not stated.

Sources

Hicks, T.G., and Wierwille, W.W. Comparison of five mental workload assessment procedures in a moving-base driving simulator. *Human Factors* 21: 129–143, 1979.
Masline, P.J. A comparison of the sensitivity of interval scale psychometric techniques in the assessment of subjective mental workload. Unpublished master's thesis, University of Dayton, Dayton, OH, 1986.

2.3.4.5 Hart and Bortolussi Rating Scale

General description – Hart and Bortolussi (1984) used a single rating scale to estimate workload. The scale units were 1 to 100 with 1 being low workload and 100 being high workload.

Strengths and limitations – The workload ratings significantly varied across flight segments with take-off and landing having higher workload than climb or cruise. The workload ratings were significantly correlated to ratings of stress (+0.75) and effort (+0.68). These results were based on data from 12 instrument-rated pilots reviewing a list of 163 events.

Moray et al. (1991) used the same rating scale but numbered the scale from 1 to 10 rather than from 1 to 100. This measure was significantly related to time pressure but not to knowledge or their interaction.

Data requirements – The participants need only the end points of the scale.
Thresholds – 1 equals low workload; 100 equals high workload.

Sources

Hart, S.A., and Bortolussi, M.R. Pilot errors as a source of workload. *Human Factors* 25(5): 545–556, 1984.
Moray, N., Dessouky, M.I., Kijowski, B.A., and Adapathya, R.S. Strategic behavior, workload, and performance in task scheduling. *Human Factors* 33(6): 607–629, 1991.

2.3.4.6 Instantaneous Self Assessment (ISA)

General description – The Instantaneous Self Assessment (ISA) is a five-point rating scale (see Table 2.26) that was originally developed in the United Kingdom to evaluate workload of Air Traffic Controllers. ISA has since been

TABLE 2.26

Instantaneous Self Assessment

ISA Button Number	Color	Legend	Definition
5	Red	VERY HIGH	Workload level is too demanding and unsustainable, even for a short period of time
4	Yellow	HIGH	Workload level is uncomfortably high, although it can be sustained for a short period of time
3	White	FAIR	Workload level is sustainable and comfortable
2	Green	LOW	Workload level is low, with occasional periods of inactivity. Operator has considerable spare capacity and is relaxed
1	Blue	VERY LOW	Workload level is too low. Operator is resting or not contributing to crew tasks

applied to evaluating workload of Joint Strike Fighter pilots. Online access to workload ratings was added to ISA and the resultant system renamed Eurocontrol Recording and Graphical display On-line (ERGO) (Hering and Coatleven, 1994).

Strengths and limitations – Hering and Coatleven (1996) stated that the ISA has been used in Air Traffic Control simulations since 1993. Lamoureux (1999) compared 81 categories of aircraft relationships in Air Traffic Control and predicted actual ISA versus ISA subjective workload ratings. The predictions were 73% accurate. Harmer (1998) adapted ISA for measurement of multi-crew workload.

Castle and Leggatt (2002) performed a laboratory experiment to compare the workload estimates from three rating scales: ISA, NASA TLX, and the Bedford Workload Scale. They asked 16 pilots and 16 nonpilots to rate their workload using each of these three workload scales while performing the Multiple Attribute Task Battery. As a control, participants also performed the task battery without rating their workload. Finally, participants were asked to complete a face validity questionnaire. Average ratings for the 11 scales on the questionnaire were between 4 and 6 on a scale of 1 to 7 (7 being the highest positive rating). This was comparable to the other two workload measures.

There were, however, significant differences between the two groups. The nonpilots rated the ISA to be significantly more professional in appearance and the pilots rated the ISA to be significantly more reliable. ISA was not sensitive to differences between pilots and nonpilots in the performance of a task battery designed to simulate flying a fixed-wing aircraft. The correlation between ISA and the Bedford Workload Scale was +0.49 and the NASA TLX was +0.55. The correlation with ratings of observers with ISA ratings was +0.80. The correlation with task loading on the Multiple Attribute Task Battery was highest for the ISA (+0.82) and lower for the NASA TLX (+0.57) and the Bedford Workload Scale (+0.53). There were no significant correlations between ISA rating and performance. Nor were there significant effects on performance whether or not the ISA rating was given. That was also true for the NASA TLX and Bedford Workload Scale. Internal consistency as measured by Cronbach's alpha varied between 0.43 and 0.78 for participants and 0.64 to 0.81 for observers. Retest reliability for the same task performed two weeks later was +0.84. Instant ratings were reported to be more consistent than ratings made two minutes after the task. In addition, Leggatt (2005) reported that construct validity for ISA was high (0.82).

Tattersall and Foord (1996) in a laboratory study using a tracking task reported that tracking task performance decreased when ISA responses were made and therefore warning of its intrusiveness on primary task performance.

Data requirements – Use of the standard rating scale. An automated system has been developed for online use. It is the Eurocontrol Recording and Graphical display On-line (ERGO)) (Hering and Coatleven, 1996).

Thresholds – 1 to 5. Sturrock and Fairburn (2005) provided the following red line values:

Development/risk reduction workload assessments

1–4 acceptable

5 investigate further

Sources

Castle, H., and Leggatt, H. *Instantaneous Self Assessment (ISA) – Validity & Reliability (JS 14865 Issue 1)*. Bristol, United Kingdom: BAE Systems, November 2002.

Harmer, S. Multi-crew workload measurement for Nimrod MRA4. Proceedings of the North Atlantic Treaty Organization Research and Technology Organization Meeting 4 (RTO-MP-4, AC/23(HFM)TP/2), 8-1–8-6, April 1998.

Hering, H., and Coatleven, G. ERGO (Version 1) for Instantaneous Self Assessment of Workload (EEC Note No. 24/94). Brussels, Belgium: EUROCONTROL Agency, April 1994.

Hering, H., and Coatleven, G. ERGO (Version 2) for Instantaneous Self Assessment of Workload in a real-time ATC simulation environment (EEC Report No. 10/96). Bruxelles, Belgium: EUROCONTROL Agency, April 1996.

Lamoureux, T. The influence of aircraft proximity data on the subjective mental workload of controllers in the air traffic control task. *Ergonomics* 42(11): 1482–1491, 1999.

Leggatt, A. Validation of the ISA (Instantaneous Self Assessment) subjective workload tool. Proceedings of the International Conference on Contemporary Ergonomics (CE2005), 74–78, April 2005.

Sturrock, F., and Fairburn, C. Measuring pilot workload in single and multi-crew aircraft. Proceedings of the International Conference on Contemporary Ergonomics (CE2005), 588–592, April 2005.

Tattersall, A.J., and Foord, P.S. An experimental evaluation of instantaneous self-assessment as a measure of workload. *Ergonomics* 39(5): 740–748, 1996.

2.3.4.7 McDonnell Rating Scale

General description – The McDonnell rating scale (see Figure 2.18) is a 10-point scale requiring a pilot to rate workload based on the attentional demands of a task.

Strengths and limitations – Van de Graaff (1987) reported significant differences in workload among various flight approach segments and crew conditions. Intersubject variability among McDonnell ratings was less than that among SWAT ratings.

Data requirements – Not stated.

Thresholds – Not stated.

Sources

McDonnell, J.D. *Pilot Rating Techniques for the Estimation and Evaluation of Handling Qualities (AFFDL-TR-68-76)*. Wright-Patterson Air Force Base, TX: Air Force Flight Dynamics Laboratory, 1968.

van de Graaff, R.C. *An In-Flight Investigation of Workload Assessment Techniques for Civil Aircraft Operations (NLR-TR-87119U)*. Amsterdam, the Netherlands: National Aerospace Laboratory, 1987.

2.3.4.8 Overall Workload Scale

General description – The Overall Workload (OW) Scale is a bipolar scale ("low" on the left; "high" on the right) requiring participants to provide a single workload rating on a horizontal line divided into 20 equal intervals.

Strengths and limitations – OW has been used in assessing workload in mobile air defense missile systems (Hill et al., 1988), remotely piloted vehicle systems (Byers et al., 1988), helicopter simulators (Iavecchia et al., 1989), and laboratories (Harris et al., 1995). The scale can be used retrospectively or prospectively (Eggleston and Quinn, 1984).

Hill et al. (1992) reported that OW was consistently more sensitive to workload and had greater operator acceptance than the Modified Cooper-Harper rating scale or the SWAT. Harris et al. (1992) reported that the OW Scale has been sensitive across tasks, systems, and environments.

Anthony and Biers (1997), however, found no difference between OW and SWAT ratings. Their participants were 48 introductory psychology students performing a memory recall task. The OW scale is less valid and reliable

Controllable Capable of being controlled or managed in context of mission, with available pilot attention	**Acceptable** May have deficiencies which warrant improvement, but adequate for mission. Pilot compensation, if required to achieve acceptable performance, is feasible.	**Satisfactory** Meets all requirements and expectations; good enough without improvement. Clearly adequate for mission.	Excellent, Highly desirable	A1
			Good, pleasant, well behaved	A2
			Fair. Some mildly unpleasant characteristics. Good enough for mission without improvement.	A3
		Unsatisfactory Reluctantly acceptable. Deficiencies which warrant improvement. Performance adequate for mission with feasible pilot compensation.	Some minor but annoying deficiencies. Improvement is requested. Effect on performance is easily compensated for by pilot.	A4
			Moderately objectionable deficiencies. Improvement is needed. Reasonable performance requires considerable pilot compensation.	A5
			Very objectionable deficiencies. Major improvements are needed. Requires best available pilot compensation to achieve acceptable performance.	A6
	Unacceptable Deficiencies which require mandatory improvement. Inadequate performance for mission, even with maximum feasible pilot compensation.		Major deficiencies which require mandatory improvement for acceptance. Controllable. Performance inadequate for mission, or pilot compensation required for minimum acceptable performance in mission is too high.	U7
			Controllable with difficulty. Requires substantial pilot skill and attention to retain control and continue mission.	U8
			Marginally controllable in mission. Requires maximum available pilot skill and attention to retain control	U9
Uncontrollable Control will be lost during some portion of the mission.			**Uncontrollable in Mission**	U10

FIGURE 2.18
McDonnell rating scale. (From McDonnell, 1968, p. 7.)

than NASA TLX or Analytical Hierarchy Process (AHP) ratings (Vidulich and Tsang, 1987).

Hall (2009) used a rating of 1 (very low) through 7 (very high) as a measure of overall workload. He called the measure the continuous subjective workload assessment technique (C-SWAT). It was presented visually to his participants every five minutes in a 25-minute scenario. It was associated with significant differences between baseline and enhanced command and control displays.

Data requirements – Not stated.
Thresholds – Not stated.

Sources

Anthony, C.R., and Biers, D.W. Unidimensional versus multidimensional workload scales and effect of number of rating scale categories. Proceedings of the Human Factors and Ergonomics Society 41st Annual Meeting, 1084–1088, 1997.

Byers, J.C., Bittner, A.C., Hill, S.G., Zaklad, A.L., and Christ, R.E. Workload assessment of a remotely piloted vehicle (RPV) system. Proceedings of the Human Factors Society 32nd Annual Meeting, 1145–1149, 1988.

Eggleston, R.G., and Quinn, T.J. A preliminary evaluation of a projective workload assessment procedure. Proceedings of the Human Factors Society 28th Annual Meeting, 695–699, 1984.

Hall, D.S. Raptor: An empirical evaluation of an ecological interface designed to increase warfighter cognitive performance. Master's Thesis. Monterey, CA: Naval Postgraduate School, June 2009.

Harris, W.C., Hancock, P.A., Arthur, E.J., and Caird, J.K. Performance, workload, and fatigue changes associated with automation. *International Journal of Aviation Psychology* 5(2): 169–185, 1995.

Harris, R.M., Hill, S.G., Lysaght, R.J., and Christ, R.E. *Handbook for Operating the OWLKNEST Technology (ARI Research Note 92-49)*. Alexandria, VA: United States Army Research Institute for the Behavioral and Social Sciences, 1992.

Hill, S.G., Iavecchia, H.P., Byers, J.C., Bittner, A.C., Zaklad, A.L., and Christ, R.E. Comparison of four subjective workload rating scales. *Human Factors* 34: 429–439, 1992.

Hill, S.G., Zaklad, A.L., Bittner, A.C., Byers, J.C., and Christ, R.E. Workload assessment of a mobile air defense missile system. Proceedings of the Human Factors Society 32nd Annual Meeting, 1068–1072, 1988.

Iavecchia, H.P., Linton, P.M., and Byers, J.C. Operator workload in the UH-60A Black Hawk crew results vs. TAWL model predictions. Proceedings of the Human Factors Society 33rd Annual Meeting, 1481–1485, 1989.

Vidulich, M.A., and Tsang, P.S. Absolute magnitude estimation and relative judgment approaches to subjective workload assessment. Proceedings of the Human Factors Society 31st Annual Meeting, 1057–1061, 1987.

2.3.4.9 Pilot Objective/Subjective Workload Assessment Technique

General description – The Pilot Objective/Subjective Workload Assessment Technique (POSWAT) is a 10-point subjective scale (see Table 2.27) developed at the Federal Aviation Administration's Technical Center (Stein, 1984). The scale is a Modified Cooper-Harper Rating Scale but does not include the binary decision tree that is characteristic of the Cooper-Harper Rating Scale. It does, however, divide workload into five categories: low, minimal, moderate, considerable, and excessive. Like the Cooper-Harper Rating Scale, the lowest three levels (1 through 3) are grouped into a low category. A similar scale, the Air Traffic Workload Input Technique (ATWIT), has been developed for Air Traffic Controllers (Porterfield, 1997). Another version is called the Workload Assessment Keypad (WAK) requiring a workload rating from 1 (very low) to 10 (very high) every two minutes (Zingale et al., 2010).

Strengths and limitations – The immediate predecessor of POSWAT was the Workload Rating System. It consisted of a workload entry device with an array of 10 pushbuttons. Each pushbutton corresponded to a rating from 1 (very easy) to 10 (very hard). The scale was sensitive to changes in flight control stability (Rehman et al., 1983). It also generally decreased as experience increased (Mallery, 1987).

Stein (1984) reported that POSWAT ratings significantly differentiated experienced and novice pilots and high (initial and final approach) and low

TABLE 2.27

POSWAT (Mallery and Maresh, 1987, p. 655)

Pilot Workload	Workload	Characteristics
1	Little or none	Any tasks completed immediately. Nominal control inputs, no direct communications. No required planning. No chance of any deviations.
2	Minimal	All tasks easily accomplished. No chance of deviation.
3	Minimal	All tasks accomplished. Minimal chance of deviation.
4	Moderate	All tasks accomplished. Tasks are prioritized. Minimal chance of deviation.
5	Moderate	All tasks are accomplished. Tasks are prioritized. Moderate chance of deviation.
6	Moderate	All tasks are accomplished. Tasks are prioritized. Considerable chance of deviation.
7	Considerable	Almost all tasks accomplished. Lowest priority tasks dropped. Considerable chance of deviation.
8	Considerable	Most tasks accomplished. Lower priority tasks dropped. Considerable chance of deviation.
9	Excessive	Only high priority tasks accomplished. Chance of error or major deviations. Second pilot desired for flight.
10	Excessive	Only highest priority tasks (safety of flight) tasks accomplished. Errors or frequent deviations occur. Second pilot needed for safe flight.

(en route) flight segments. There was also a significant learning effect: workload ratings were significantly higher on the first than on the second flight. Although the POSWAT scale was sensitive to manipulations of pilot experience level for flights in a light aircraft and in a simulator, the scale was cumbersome. Seven dimensions (workload, communications, control inputs, planning, "deviations," error, and pilot complement) are combined on one scale. Further, the number of ranks on the ordinal scale is confusing since there are both five and 10 levels.

Rehman et al. (1983) obtained POSWAT ratings once per minute. These investigators found that pilots reliably reported workload differences in a tracking task on a simple 10-point non-adjectival scale. Therefore, the cumbersome structure of the POSWAT scale may not be necessary. Zingale et al. (2010) reported significant interaction of Air Traffic Control conditions and interval for WAK. However, there was no significant effect of type of Air Traffic Control workstation. Hah et al. (2010) reported significant differences in WAK scores for variations in human machine interfaces to Data Communications as well as for the percent equipage (0%, 10%, 50%, or 100%). There was no significant effect, however, of a partial or system-wide failure on WAK.

Data requirements – Stein (1984) suggested not analyzing POSWAT ratings for short flight segments if the ratings are given at one-minute intervals.

Thresholds – Not stated.

Sources

Hah, S., Willems, B., and Schultz, K. The evaluation of Data Communication for the Future Air Traffic Control System (NextGen). Proceedings of the Human Factors and Ergonomics Society 54th Annual Meeting, 99–103, 2010.

Mallery, C.J. The effect of experience on subjective ratings for aircraft and simulator workload during IFR flight. Proceedings of the Human Factors Society 31st Annual Meeting, 838–841, 1987.

Mallery, C.J., and Maresh, J.L. Comparison of POSWAT ratings for aircraft and simulator workload. Proceedings of the 4th International Symposium on Aviation Psychology, 644–650, 1987.

Porterfield, D.H. Evaluating controller communication time as a measure of workload. *The International Journal of Aviation Psychology* 7(2): 171–182, 1997.

Rehman, J.T., Stein, E.S., and Rosenberg, B.L. Subjective pilot workload assessment. *Human Factors* 25(3): 297–307, 1983.

Stein, E.S. *The Measurement of Pilot Performance: A Master-Journeyman Approach (DOT/FAA/CT-83/15)*. Atlantic City, NJ: Federal Aviation Administration Technical Center, May 1984.

Zingale, C.M., Willems, B., and Ross, J.M. Air Traffic Controller workstation enhancements for managing high traffic levels and delegated aircraft procedures. Proceedings of the Human Factors and Ergonomics Society 54th Annual Meeting, 11–15, 2010.

2.3.4.10 Utilization

General description – Utilization (p) is the probability of the operator being in a busy status (Her and Hwang, 1989).

Strengths and limitations – Utilization has been a useful measure of workload in continuous process tasks (e.g., milling, drilling, system controlling, loading, and equipment setting). It accounts for both arrival time of work in a queue and service time on that work.

Data requirements – A queuing process must be in place.

Thresholds – Minimum value is 0, maximum value is 1. High workload is associated with the maximum value.

Source

Her, C., and Hwang, S. Application of queuing theory to quantify information workload in supervisory control systems. *International Journal of Industrial Ergonomics* 4: 51–60, 1989.

2.3.5 Task-Analysis Based Subjective Workload Measures

These measures break the tasks into subtasks and subtask requirements for workload evaluation. Examples include: Arbeitswissenshaftliches Erhebungsverfahren zur Tatigkeitsanalyze (Section 2.3.5.1), Computerized Rapid Analysis of Workload (Section 2.3.5.2), McCracken-Aldrich Technique (Section 2.3.5.3), Task Analysis Workload (Section 2.3.5.4), and Zachary/Zaklad Cognitive Analysis (Section 2.3.5.5).

2.3.5.1 *Arbeitswissenshaftliches Erhebungsverfahren zur Tatigkeitsanalyze*

General description – Arbeitswissenschaftliches Erhebungsverfahren zur Tatigkeitsanalyze (AET) was developed in Germany to measure workload. AET has three parts: (1) work system analysis, which rates the "type and properties of work objects, the equipment to be used as well as physical social and organizational work environment" (North and Klaus, 1980, p. 788) on both nominal and ordinal scales; (2) task analysis, which uses a 31-item ordinal scale to rate "material work objects, abstract (immaterial) work objects, and man-related tasks" (p. 788); and (3) job-demand analysis, which is used to evaluate the conditions under which the job is performed. "The 216 items of the AET are rated on nominal or ordinal scales using five codes as indicated for each item: frequency, importance, duration, alternative and special (intensity) code" (p. 790).

Strengths and limitations – AET has been used in over 2,000 analyses of both manufacturing and management jobs.

Data requirements – Profile analysis is used to analyze the job workload. Cluster analysis is used to identify elements of jobs that "have a high degree of natural association among one another" (p. 790). Multivariate statistics are used for "placement, training, and job classification" (p. 790).

Thresholds – Not stated.

Source

North, R.A., and Klaus, J. Ergonomics methodology – An obstacle or promoter for the implementation of ergonomics in industrial practice? *Ergonomics* 23(8): 781–795, 1980.

2.3.5.2 Computerized Rapid Analysis of Workload

General description – The Computerized Rapid Analysis of Workload (CRAWL) is a computer program that helps designers predict workload in systems being designed. CRAWL inputs are mission timelines and task descriptions. Tasks are described in terms of cognitive, psychomotor, auditory, and visual demands.

Strengths and limitations – Bateman and Thompson (1986) reported increases in CRAWL ratings as task difficulty increased. Vickroy (1988) reported similar results as air turbulence increased.

Data requirements – The mission timeline must provide detailed second-by-second descriptions of the aircraft status.

Thresholds – Not stated.

Sources

Bateman, R.P., and Thompson, M.W. Correlation of predicted workload with actual workload measured using the Subjective Workload Assessment Technique. Proceedings of the SAE AeroTech Conference, 1986.

Vickroy, C.C. *Workload Prediction Validation Study: The Verification of CRAWL Predictions.* Wichita, KS: Boeing Military Airplane Company, 1988.

2.3.5.3 McCracken-Aldrich Technique

General description – The McCracken-Aldrich Technique was developed to identify workload associated with flight control, flight support, and mission-related activities (McCracken and Aldrich, 1984).

Strength and limitations – The technique may require months of preparation to use. It has been useful in assessing workload in early system design stages.

Data requirements – A mission must be decomposed into segments, functions, and performance elements (e.g., tasks). Subject Matter Experts rate workload (from 1 to 7) for each performance element. A FORTRAN programmer is required to generate the resulting scenario timeline.

Thresholds – Not stated.

Source

McCracken, J.H., and Aldrich, T.B. *Analysis of Selected LHX Mission Functions: Implications for Operator Workload and System Automation Goals (TNA ASI 479-24-84)*. Fort Rucker, AL: Anacapa Sciences, 1984.

2.3.5.4 Task Analysis Workload

General description – The Task Analysis/Workload (TAWL) technique requires missions to be decomposed into phases, segments, functions, and tasks. For each task a Subject Matter Expert rates the workload on a scale of 1 to 7. The tasks are combined into a scenario timeline and workload estimated for each point on the timeline.

Strengths and limitations – TAWL is sensitive to task workload but requires about six months to develop (Harris et al., 1992). It has been used to identify workload in helicopters (Szabo and Bierbaum, 1986).

$$p = b_0 + b_1 N + b_2 S$$

where:
 p = utilization
 b = intercept determined from regression analysis
 b_1 = slope determined from regression analysis
 N = number of information types
 S = quantity of information in an information type

Hamilton and Cross (1993), based on seven task conditions, performance of 20 AH-64 aviators, and two analysts, reported significant correlations (+0.89 and +0.99) between the measures predicted by the TAWL model and the actual data.

Data requirements – A detailed task analysis is required. Then Subject Matter Experts must rate the workload of each task on six channels (auditory, cognitive, kinesthetic, psychomotor, visual, and visual-aided). A PC compatible system is required to run the TAWL software. A user's guide (Hamilton et al., 1991) is available.

Thresholds – Not stated.

Sources

Hamilton, D.B., Bierbaum, C.R., and Fulford, L.A. *Task Analysis/Workload (TAWL) User's Guide – Version 4.0 (ASI 690-330-90)*. Fort Rucker, AL: Anacapa Sciences, 1991.

Hamilton, D.B., and Cross, K.C. *Preliminary Validation of the Task Analysis/Workload Methodology (ARI RN92-18)*. Alexandria, VA: Army Research Institute for the Behavioral and Social Sciences, 1993.

Harris, R.M., Hill, S.G., Lysaght, R.J., and Christ, R.E. *Handbook for Operating the OWL & NEST Technology (ARI Research Note 92-49)*. Alexandria, VA: United States Army Research Institute for the Behavioral and Social Sciences, 1992.

Szabo, S.M., and Bierbaum, C.R. *A Comprehensive Task Analysis of the AH-64 Mission with Crew Workload Estimates and Preliminary Decision Rules for Developing an AH-64 Workload Prediction Model (ASI 678-204-86[B])*. Fort Rucker, AL: Anacapa Sciences, 1986.

2.3.5.5 Zachary/Zaklad Cognitive Analysis

General Description – The Zachary/Zaklad Cognitive Analysis Technique requires both operational Subject Matter Experts and cognitive scientists to identify operator strategies for performing all tasks listed in a detailed cognitive mission task analysis. A second group of Subject Matter Experts then rates, using 13 subscales, workload associated with performing each task.

Strengths and limitations – The method has only been applied in two evaluations, one for the P-3 aircraft (Zaklad et al., 1982), the other for F/A-18 aircraft (Zachary et al., 1987; Zaklad et al., 1987).

Data requirements – A detailed cognitive mission timeline must be constructed. Two separate groups of Subject Matter Experts are required: one to develop the timeline, the other to rate the associated workload.

Thresholds – Not stated.

Sources

Zachary, W., Zaklad, A., and Davis, D. A cognitive approach to multisensor correlation in an advanced tactical environment. Technical Proceedings of the 1987 Tri-Service Data Fusion Symposium, 438–462, 1987.

Zaklad, A.L., Deimler, J.D., Iavecchia, H.P., and Stokes, J. Multisensor correlation and TACCO workload in representative ASW and ASUW environments (Analytics Tech Report-1753A) Willow Grove, PA: Analytics, 1982.

Zaklad, A., Zachary, W., and Davis, D. A cognitive model of multisensor correlation in an advanced aircraft environment. Proceedings of the 4th Midcentral Ergonomics/Human Factors Conference, 59–65, 1987.

2.4 Simulation of Workload

Several digital models have been used to evaluate workload. These include: (1) Null Operation System Simulation (NOSS), (2) SAINT (Buck, 1979), Modified Petri Nets (MPN) (White et al., 1986), (3) task analyses (Bierbaum and Hamilton, 1990), and (4) Workload Differential Model (Ntuen and Watson, 1996).

In addition, Riley (1989) described the Workload Index (W/INDEX), a simulation which enables designers to compare alternate physical layouts and interface technologies, use of automation, and sequence of tasking. The designer inputs the system design concept and assigns each to an interface channel (i.e., visual, auditory, manual, verbal) and a task timeline. W/INDEX then predicts workload. A Micro Saint model of the workload in a simulated air-to-ground combat mission did not predict workload as measured by SWAT (See and Vidulich, 1997).

More recently a number of workload models have been developed based on the Improved Performance Research Integration Tool (IMPRINT). An example for driver workload is reported by Kandemir et al. (2018). Rusnock and Geiger (2017) presented a model for military operations using UAVs.

Sources

Bierbaum, C.R., and Hamilton, D.B. Task analysis and workload prediction model of the MH-60K mission and a comparison with UH-60A workload predictions; Volume III: Appendices H through N (ARI Research Note 91-02). Alexandria, VA: U.S. Army Research Institute for the Behavioral and Social Sciences, October 1990.

Buck, J. Workload estimation through simulation paper presented at the Workload Program of the Indiana Chapter of the Human Factors Society, Crawfordsville, Indiana, March 31, 1979.

Kandemir, C., Handley, H.A.H., and Thompson, D. A workload model to evaluate distracters and driver's aids. *International Journal of Industrial Ergonomics*, 16: 18–36, 2018.

Ntuen, C.A., and Watson, A.R. Workload prediction as a function of system complexity. Proceedings of the 3rd Annual Symposium on Human Interaction with Complex Systems, 96–100, 1996.

Riley, V. W/INDEX: A crew workload prediction tool. Proceedings of the 5th International Symposium on Aviation Psychology, 832–837, 1989.

Rusnock, C.F., and Geiger, C.D. Simulation-based evaluation of adaptive automation revoking strategies on cognitive workload and situation awareness. *IEEE Transactions on Human-Machine Systems* 47(6): 927–938, 2017.

See, J.E., and Vidulich, M.A. Assessment of computer modeling of operator mental workload during target acquisition. Proceedings of the Human Factors and Ergonomics Society 41st Annual Meeting, 1303–1307, 1997.

White, S.A., MacKinnon, D.P., and Lyman, J. *Modified Petri Net Modal Sensitivity to Workload Manipulations (NASA-CR-177030)*. Moffett Field, CA: NASA Ames Research Center, 1986.

2.5 Dissociation of Workload and Performance

Wickens and Yeh (1983) presented a theory of dissociation between subjective measures of workload and performance. The theory proposes that the subjective measures are determined by the number of tasks and performance by task difficulty. In subsequent work, Yeh and Wickens (1985, 1988) identified five conditions in which the relationship between performance and subjective measures of workload imply different effects on workload. In the first condition, termed motivation, performance improves and subjective ratings of workload increase (see Figure 2.19). In the second condition, which Yeh and Wickens termed underload, as demand increases, performance remains the same but subjective measures of workload increase. In this condition, performance implies that workload remains constant while the subjective measures imply increased workload. In the third condition, resource-limited tasks, as the amount of invested resources increases, performance degrades and subjective workload measures increase. However, the proportion of the change in subjective ratings is greater than the proportion of the change in performance. In this condition, performance implies that workload increases somewhat while subjective workload measures imply that workload increases greatly. In the fourth condition, comparison of dual-task configurations with different degrees of competition for common resources, as demand for common resources increases, performance degrades and subjective workload measures increase. This time, however, the change in performance is greater than the change in subjective ratings. In this condition, performance implies that workload increases greatly, while subjective workload measures suggest that workload increases slightly. In the fifth condition, which Yeh and Wickens termed overload, as demand increases, performance degrades while subjective measures remain unchanged. In this

Human Workload

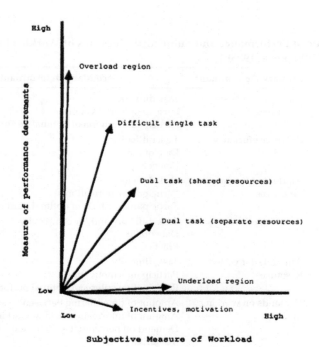

FIGURE 2.19
Dissociations between performance and subjective measures of workload as predicted by theory (Adapted from Yeh and Wickens, 1988, p. 115).

condition, performance implies that workload increases while the subjective measures imply that workload remains the same.

Yeh and Wickens defined these differences in implied workload as dissociation and they suggested that dissociation occurs because, in these five conditions, different factors determine performance and subjective workload measures. These factors are listed in Table 33. As can be seen from Table 2.28, there is only one factor common to both performance and subjective measures of workload: amount of invested resources. However, there are different resource dichotomies. For example, Yeh and Wickens (1988) defined four types of resources: (1) perceptual/central versus response stage, (2) verbal versus spatial code, (3) visual versus auditory input modality, and (4) manual versus speech output modality.

The importance of these dichotomies has been supported in many experiments. For example, Wickens and Liu (1988) reported greater tracking error when a one-dimensional compensatory tracking task was time-shared with a spatial decision task (same code: spatial) than with a verbal decision task (different codes: spatial and verbal). Tracking error was also greater when the response to the time-shared decision task was manual (same modality: manual) rather than verbal (different modality: manual versus speech).

TABLE 2.28

Determinants of Performance and Subjective Measures of Workload (Adapted from Yeh and Wickens, 1988)

Measure	Primary Determinant	Secondary Determinant
Single-task performance	Amount of invested Resources	Task difficulty Participant's motivation Participative criteria of optimal performance
	Resource efficiency	Task difficulty Data quality Practice
Dual-task performance	Amount of invested Resources	Task difficulty Participant's motivation Participative criteria of optimal performance
	Resource efficiency	Task difficulty and/or complexity Data quality Practice
Subjective Workload	Amount of invested Resources	Task difficulty Participant's motivation Participative criteria of optimal performance
	Demands on working Memory	Amount of time sharing between tasks Amount of information held in working memory Demand on perceptual and/or central processing Resources

Derrick (1985) analyzed performance on 18 computer-based tasks and a global subjective estimate for workload on each task. The tasks were performed in four configurations: (1) single easy, (2) single hard, (3) dual with the same task, and (4) dual with different tasks. He concluded: "If a task was increased in difficulty by adding to the perceptual and central processing resource load (and thus performance declined), people rated the workload as higher than for the same task with lower perceptual and central processing resource demands. However, tasks made more difficult by increasing the resource demands of responding (leading to worse performance) did not produce increased workload ratings" (p. 1022). In addition, "performance in the Dual with the Same Task configuration was worse than in the Dual with Different Tasks configuration, but the workload ratings were essentially equivalent for these two conditions" (p. 1022).

Derrick summarized his findings as well as those of Yeh and Wickens (1988) in Table 2.29. The values in the cells indicate the weight associated with each portion of the dissociation effect.

The relationship between the factors that determine performance and subjective measures of workload and the resource dichotomies is complex. Greater resource supply improves performance of resource-limited tasks but not of data-limited tasks. For dual tasks, performance degrades when the tasks compete for common resources.

The importance of accurate performance and workload measurement during system evaluation cannot be overstated. As Derrick (1985) stated:

TABLE 2.29

A Theory of Dissociation

	Sources	Performance Decreases	Subjective Difficulty Increases
1	Increased single task difficulty	4	3
	Perceptual/cognitive	2	2
	Response	2	1
2	Concurrent task demand	3	4
	Same resources	2	2
	Different resources	1	2

"A workload practitioner who relies on these [subjective] ratings [of workload] rather than performance will be biased to choose a nonoptimal system that requires operators to perform just one task rather than the system that demands dual task performance" (p. 1023). "The practitioner who relies solely on subjective data may likely favor the system that has serious performance limitations, especially under emergency conditions" (p. 1024). "A system design option that increases control demands, and ultimately degrades performance under high workload conditions, will not be rated as a problem in the normal single task workload evaluations" (p. 1024). Therefore, the potential for inaccurate interpretation of the results is worrisome.

Gawron made an extensive search, which was made of Defense Technical Information Center (DTIC) reports and refereed journals, to identify publications in which both task performance and subjective measures of workload were reported. Only 11 publications meeting this criterion were found (see Table 2.30). The publications spanned 11 years from 1978 to 1996. Eight were performed in flight simulators, one in a centrifuge, and two in a laboratory using a tracking task. The type of facility, mission, duration, number of trials, and type of participants in each of these publications are also described in Table 2.30. As can be seen from the table, missions varied from airdrop (1) to research (2) to combat (3) with half being transport missions (5). The duration of trials also varied widely from 1 to 200 minutes with up to 100 replications. Each publication reported data based on at least six participants with two publications reporting data from 48 participants. Half the studies collected data from only male participants.

The high and low workload conditions, task, performance metric, and subjective metric for each data point gleaned from these publications are presented in Table 2.31. The workload conditions covered the entire range from physical (flight acceleration forces), mental effort (inexperience, different responses to the same cues, difficult navigation problems, dual task, and no autopilot), psychological stress (occurrence of malfunctions, high winds), and time stress (high target rates). Performance metrics included both speed (1 publication) and error (10 publications) measures. The subjective workload metrics were also diverse: four publications reported a workload rating,

TABLE 2.30
Description of Studies Reviewed

Study	Reference	Facility	Mission	Duration	Trials	Participants	Laboratory
1	Albery (1989)	Dynamic environment simulator (centrifuge)	Combat	60 seconds	24 or 12 per participant	9 male military personnel	Air Force Research Laboratory
2	Casali and Wierwille (1983)	Singer/Link GAT-1B simulator	Transport	Not stated	3 per participant	29 males and 1 female civilian pilot	Virginia Polytechnic Institute and State University
3	Casali and Wierwille (1984)	Singer/Link GAT-1B simulator	Transport	12 minutes	3 per participant	48 male pilots	Virginia Polytechnic Institute and State University
4	Hancock (1996)	1-D compensatory tracking	Research	2 minutes	100 per participant	6 right-handed male university volunteers	University of Minnesota
5	Kramer et al. (1987)	ILLIMAC fixed-based flight simulator	Transport	45 minutes	4 per participant	7 right-handed male student pilots	University of Illinois
6	Madero et al. (1979)	Air force flight dynamics Laboratory multicrew simulator	Air drop	Not stated	3 per crew	8 C-130 crews made up of pilot, copilot, and loadmaster	Air Force Research Laboratory
7	Stein (1984)	Singer/Link general aviation trainer (simulated Cessna 421)	Transport	35 minutes	2 per participant	12 air transport commercial pilots and 12 recently qualified instrument pilots	FAA Technical Center

(Continued)

TABLE 2.30 (CONTINUED)
Description of Studies Reviewed

Study	Reference	Facility	Mission	Duration	Trials	Participants	Laboratory
8	Vidulich and Bortolussi (1988)	NASA Ames 1-Cab fixed-base simulator	Combat	1 to 1.5 hours	4	12 male Army AH-64 helicopter pilots male RAF helicopter test pilot male retired Army helicopter pilot	Air Force Research Laboratory
9	Vidulich and Wickens (1985, 1986)	2-D compensatory tracking and Sternberg task	Research	Not stated	14 per participant	40 students	University of Illinois
10	Wierwille et al. (1985)	Singer/Link GAT-1B simulator	Transport	Not stated	3 per participant	48 male pilots	Virginia Polytechnic Institute and State University
11	Wolf (1978)	F-4 simulator	Combat	2 minutes	120 total	7 RF-4B aircraft pilots from Air National Guard	Air Force Research Laboratory

TABLE 2.31

Summary of Research Reporting Both Performance and Subjective Measures of Workload

Study	Data Point	High Workload	Low Workload	Task	Performance Metric	Subjective Workload Metric
1	1	3.75G acceleration	1.4G acceleration	Compensatory tracking	Tracking error	SWAT
1	2	100 dBA noise	40 dBA noise	Compensatory tracking	Tracking error	SWAT
2	3	One call sign every 2 seconds on average; nontarget call signs were permutations	One call sign every 12 seconds on average; nontarget call signs were permutations	Control aircraft	Z-scores errors of omission in communications	Modified Cooper-Harper rating scale
2	4	One call sign every 2 seconds on average; nontarget call signs were permutations	One call sign every 12 seconds on average; nontarget call signs were permutations	Control aircraft	Z-scores errors of commission in communications	Modified Cooper-Harper rating scale
2	5	One call sign every 2 seconds on average; nontarget call signs were permutations	One call sign every 12 seconds on average; nontarget call signs were permutations	Control aircraft	Z-scores communications response time	Modified Cooper-Harper rating scale
2	6	One call sign every 2 seconds on average; nontarget call signs were permutations	One call sign every 12 seconds on average; nontarget call signs were permutations	Control aircraft	Z-scores errors of omission in communications	Multi-descriptor scale
2	7	One call sign every 2 seconds on average; nontarget call signs were permutations	One call sign every 12 seconds on average; nontarget call signs were permutations	Control aircraft	Z-scores errors of commission in communications	Multi-descriptor scale

(*Continued*)

Human Workload

TABLE 2.31 (CONTINUED)
Summary of Research Reporting Both Performance and Subjective Measures of Workload

Study	Data Point	High Workload	Low Workload	Task	Performance Metric	Subjective Workload Metric
2	8	One call sign every 2 seconds on average; nontarget call signs were permutations	One call sign every 12 seconds on average; nontarget call signs were permutations	Control aircraft	Z-scores communications response time	Multi-descriptor scale
3	9	High icing, potential malfunction in any engine or fuel gauges, average rate of malfunctions 5 seconds	Low icing, average rate of icing hazard 1 every 50 seconds	Control aircraft	Pitch high-pass mean square	Modified Cooper-Harper rating scale
3	10	High icing, potential malfunction in any engine or fuel gauges, average rate of malfunctions 5 seconds	Low icing, average rate of icing hazard 1 every 50 seconds	Control aircraft	Roll high-pass mean square	Modified Cooper-Harper rating scale
3	11	High icing, potential malfunction in any engine or fuel gauges, average rate of malfunctions 5 seconds	Low icing, average rate of icing hazard 1 every 50 seconds	Control aircraft	Response time to hazard	Modified Cooper-Harper rating scale
3	12	High icing, potential malfunction in any engine or fuel gauges, average rate of malfunctions 5 seconds	Low icing, average rate of icing hazard 1 every 50 seconds	Control aircraft	Pitch high-pass mean square	Multi-descriptor scale
3	13	High icing, potential malfunction in any engine or fuel gauges, average rate of malfunctions 5 seconds	Low icing, average rate of icing hazard 1 every 50 seconds	Control aircraft	Roll high-pass mean square	Multi-descriptor scale

(Continued)

TABLE 2.31 (CONTINUED)

Summary of Research Reporting Both Performance and Subjective Measures of Workload

Study	Data Point	High Workload	Low Workload	Task	Performance Metric	Subjective Workload Metric
3	14	High icing, potential malfunction in any engine or fuel gauges, average rate of malfunctions 5 seconds	Low icing, average rate of icing hazard 1 every 50 seconds	Control aircraft	Response time to hazard	Multi-descriptor scale
3	15	High icing, potential malfunction in any engine or fuel gauges, average rate of malfunctions 5 seconds	Low icing, average rate of icing hazard 1 every 50 seconds	Control aircraft	Pitch high-pass mean square	Workload/Compensation/Interference/Technical effectiveness scale
3	16	High icing, potential malfunction in any engine or fuel gauges, average rate of malfunctions 5 seconds	Low icing, average rate of icing hazard 1 every 50 seconds	Control aircraft	Roll high-pass mean square	Workload/Compensation/Interference/Technical effectiveness scale
3	17	High icing, potential malfunction in any engine or fuel gauges, average rate of malfunctions 5 seconds	Low icing, average rate of icing hazard 1 every 50 seconds	Control aircraft	Response time to hazard	Workload/Compensation/Interference/Technical effectiveness scale
4	18	Initial 10 trials	final 10 trials	1-D compensatory tracking	Tracking error	NASA TLX
4	19	Initial 10 trials	final 10 trials	1-D compensatory tracking	Tracking error	SWAT
4	20	Initial 10 trials 30 days later	final 10 trials 30 days later	1-D compensatory tracking	Tracking error	NASA TLX

(*Continued*)

TABLE 2.31 (CONTINUED)
Summary of Research Reporting Both Performance and Subjective Measures of Workload

Study	Data Point	High Workload	Low Workload	Task	Performance Metric	Subjective Workload Metric
4	21	Initial 10 trials 30 days later	final 10 trials 30 days later	1-D compensatory tracking	Tracking error	SWAT
5	22	30 mph wind from 270 degrees, moderate turbulence, partial suction failure in heading indicator during approach	No wind, no turbulence, no system failure	Control aircraft	Heading deviation	Subjective workload rating
5	23	30 mph wind from degrees, moderate turbulence, partial suction failure in heading indicator during approach	No wind, no turbulence, no system failure	Control aircraft	Altitude deviation	Subjective workload rating
5	24	30 mph wind from 270 degrees, moderate turbulence, partial suction failure in heading indicator during approach	No wind, no turbulence, no system failure	Control aircraft	Glideslope deviation	Subjective workload rating
6	25	Without autopilot, without bulk data storage	With autopilot, with bulk data storage	Control aircraft	Course error during cruise segment (feet)	Subjective workload rating
6	26	Without autopilot, without bulk data storage	With autopilot, with bulk data storage	Control aircraft	Course error during CARP segment (feet)	Subjective workload rating

(Continued)

TABLE 2.31 (CONTINUED)
Summary of Research Reporting Both Performance and Subjective Measures of Workload

Study	Data Point	High Workload	Low Workload	Task	Performance Metric	Subjective Workload Metric
7	27	Journeymen (recently qualified instrument pilots)	Masters (professional, high-time pilots)	Control aircraft	Automated performance score enroute flight 1	POSWAT rating
7	28	Journeymen	Masters	Control aircraft	Automated performance score descent flight 1	POSWAT rating
7	29	Journeymen	Masters	Control aircraft	Automated performance score initial approach flight 1	POSWAT rating
7	30	Journeymen	Masters	Control aircraft	Automated performance score final approach flight 1	POSWAT rating
7	31	Journeymen	Masters	Control aircraft	Automated performance score enroute flight 2	POSWAT rating

(Continued)

Human Workload

TABLE 2.31 (CONTINUED)
Summary of Research Reporting Both Performance and Subjective Measures of Workload

Study	Data Point	High Workload	Low Workload	Task	Performance Metric	Subjective Workload Metric
7	32	Journeymen	Masters	Control aircraft	Automated performance score descent flight 2	POSWAT rating
7	33	Journeymen	Masters	Control aircraft	Automated performance score initial approach flight 2	POSWAT Rating
7	34	Journeymen	Masters	Control aircraft	Automated performance score final approach flight 2	POSWAT rating
8	35	Secondary task required speech response	Secondary task required manual response	Respond to surface-to-air missile	Response time decrement during cruise	AHP rating
8	36	Secondary task required speech response	Secondary task required manual response	Respond to surface-to-air missile	Response time decrement during hover	AHP rating
8	37	Secondary task required speech response	Secondary task required manual response	Respond to surface-to-air missile	Response time decrement during combat	AHP rating

(*Continued*)

TABLE 2.31 (CONTINUED)

Summary of Research Reporting Both Performance and Subjective Measures of Workload

Study	Data Point	High Workload	Low Workload	Task	Performance Metric	Subjective Workload Metric
9	38	Inconsistent Sternberg letters	Consistent Sternberg letters	Compensatory tracking with Sternberg	Z-scores reaction time	Z-scores task difficulty rating
10	39	Difficult navigation question, large number of numerical calculations performed, large degree of rotation of reference triangle	Easy navigation question, small number of calculations performed, small degree of rotation of reference triangle	Control aircraft	Z-scores error rate	Modified Cooper-Harper rating scale
10	40	Difficult navigation question, large number of numerical calculations performed, large degree of rotation of reference triangle	Easy navigation question, small number of calculations performed, small degree of rotation of reference triangle	Control aircraft	Z-scores reaction time	Modified Cooper-Harper rating scale
10	41	Difficult navigation question, large number of numerical calculations performed, large degree of rotation of reference triangle	Easy navigation question, small number of calculations performed, small degree of rotation of reference triangle	Control aircraft	Z-scores error rate	Workload/Compensation/ Interference/Technical Effectiveness scale
10	42	Difficult navigation question, large number of numerical calculations performed, large degree of rotation of reference triangle	Easy navigation question, small number of calculations performed, small degree of rotation of reference triangle	Control aircraft	Z-scores reaction time	Workload/Compensation/ Interference/Technical Effectiveness Scale

(*Continued*)

TABLE 2.31 (CONTINUED)
Summary of Research Reporting Both Performance and Subjective Measures of Workload

Study	Data Point	High Workload	Low Workload	Task	Performance Metric	Subjective Workload Metric
11	43	Wind gusts to 30 knots, 0.05 radian per second rate limit on aileron and stabilator deflection	No gusts, 0.5 radian per second rate limit on aileron and stabilator deflection	Control aircraft	Z-scores lateral path root mean square error	Workload Rating
11	44	Wind gusts to 30 knots, 0.05 radian per second rate limit on aileron and stabilator deflection	No gusts, 0.5 radian per second rate limit on aileron and stabilator deflection	Control aircraft	Z-scores speed root mean square error	Workload Rating
11	45	Wind gusts to 30 knots, 0.05 radian per second rate limit on aileron and stabilator deflection	No gusts, 0.5 radian per second rate limit on aileron and stabilator deflection	Control aircraft	Z-scores pitch attitude root mean square error	Workload Rating
11	46	Wind gusts to 30 knots, 0.05 radian per second rate limit on aileron and stabilator deflection	No gusts, 0.5 radian per second rate limit on aileron and stabilator deflection	Control aircraft	Z-scores vertical path root mean square error	Workload Rating
11	47	Wind gusts to 30 knots, 0.05 radian per second rate limit on aileron and stabilator deflection	No gusts, 0.5 radian per second rate limit on aileron and stabilator deflection	Control aircraft	Z-scores roll attitude root mean square error	Workload Rating

three the Modified Cooper-Harper Rating Scale, one AHP, one NASA TLX, one POSWAT, and two SWAT. Three publications reported Z-scores; seven did not. The publications were from four laboratories: Air Force Research Laboratory (3 publications/22 data points), FAA Tech Center (1/8), University of Illinois (2/4), University of Minnesota (1/4), and VPI (3/19).

All but one (point 21 in which workload decreased) of the 47 data points showed increased ratings of workload between the low and high workload conditions. All but six of the data points showed decreased performance (either increase in errors or time) in the high workload condition. The six points in which performance improved while workload increased are presented in Table 2.32. The points come from three different studies by three different research teams. They use different workload rating scales and performance metrics (although 5 of the 6 are response time). What is the same is that each of the data points involves auditory stimuli and/or responses. (Note, there was no dissociation with other performance measures for the same independent variable). These points are further categorized in Table 2.33. Five of the six points require dual-task performance in which one task requires a manual response, the other a speech response. The sixth point, data point 2, required participants to ignore an auditory stimulus to concentrate on performing a manual task.

Is the use of separate resources associated then with dissociation between performance and workload? There are two ways to test this, first against the results from similar points and second against the theoretical plot in Figure 22. There are communication requirements in data points 3 through 8 and 35 through 37. Why did dissociation not occur in data points 3, 4, 6, and 7? Perhaps because the performance metrics were errors and not time. In each case in which the performance metric was time and speech was required, dissociation occurred. In examining Figure 22, these data points would fall in the incentives or motivation region in which performance is enhanced but workload increased. Avoiding a missile is certainly motivating and this may have resulted in the dissociation between performance and workload for data points 34, 36, and 37. For points 5 and 8, participants were instructed to "strive to maintain adequate (specified) performance on all aspects of the primary task" (p. 630). These words may have motivated the participants.

Dissociation between performance and subjective measures of workload occurred in 7 out of 47 data points reviewed above. In five points, performance time decreased while subjective workload ratings increased. These points came from three different experiments, resulted from four different types of workload (flight acceleration forces, communications difficulty, input mode, time), used three different tasks (tracking, controlling aircraft, responding to a surface-to-air missile), and four different subjective workload metrics (SWAT, Modified Cooper-Harper Rating Scale, Multi-descriptor scale, and AHP). This diversity supports the existence of the dissociation phenomenon and suggests the need for guidelines in the interpretation of workload data. As a start, Yeh and Wickens (1988) suggest that performance

Human Workload

TABLE 2.32
Points in Which Workload Increased and Performance Improved

Data Point	High Workload	Low Workload	Task	Performance Metric	Subjective Workload Metric
2	100 dBA	40 dBA	Compensatory tracking	Tracking error	SWAT
5	One call sign every 2 seconds on average; nontarget call signs were permutations	One call sign every 12 seconds on average; nontarget call signs were permutations	Control aircraft	Z-scores communications response time	Modified Cooper-Harper Rating Scale
8	One call sign every 2 seconds on average; nontarget call signs were permutations	One call sign every 12 seconds on average; nontarget call signs were permutations	Control aircraft	Z-scores communications response time	Multi-descriptor Scale
21	Initial 10 trials 30 days later	Final 10 trials 30 days later	1-D compensatory tracking	Tracking error	SWAT
35	Secondary task required speech response	Secondary task required manual response	Respond to surface-to-air missile	Response time decrement during cruise	AHP Rating
36	Secondary task required speech response	Secondary task required manual response	Respond to surface-to-air missile	Response time decrement during hover	AHP Rating
37	Secondary task required speech response	Secondary task required manual response	Respond to surface-to-air missile	Response time decrement during combat	AHP Rating

TABLE 2.33
Categorization of above Points

Data Point	Task Primary	Task Secondary	Resources Shared	Resources Separate
2	Tracking			
5	Tracking	Communicating		X
8	Tracking	Communicating		X
21	Tracking			
35	Tracking	Reaction Time		X
36	Tracking	Reaction Time		X
37	Tracking	Reaction Time		X

measures be used to provide a "direct index of the benefit of operating a system" (p. 118). Further, subjective workload measures should be used to "indicate potential performance problems that could emerge if additional demands are imposed" (p. 118).

King et al. (1989) tested the multiple resource model by examining RT and errors on four simulated flight tasks. Workload was measured using NASA-TLX. They reported mixed results: the significant decrement associated with single-to-dual-task performance was paired with increased overall workload ratings. However, the discrete tasks were unaffected by the single-to-dual tasking but workload ratings increased. Tracking degraded in the dual-task condition and workload also decreased.

Bateman et al. (1984) examined the relationship between error occurrence and SWAT ratings. They concluded that errors are dependent on task structure and workload on task difficulty and situational stress.

Thornton (1985) based on a hovercraft simulation that dissociation was most pronounced when workload increased at the beginning of the mission.

Sources

Albery, W.B. The effect of sustained acceleration and noise on workload in human operators. *Aviation, Space, and Environmental Medicine* 60(10): 943–948, 1989.
Bateman, R. P., Acton, W. H., and Crabtree, M.S. Workload and performance: Orthogonal measures. Proceedings of the Human Factors Society 28th Annual Meeting, 678–679, 1984.
Casali, J.G., and Wierwille, W.W. A comparison of rating scale, secondary task, physiological, and primary-task workload estimation techniques in a simulated flight task emphasizing communications load. *Human Factors* 25: 623–642, 1983.
Casali, J.G., and Wierwille, W.W. On the comparison of pilot perceptual workload: A comparison of assessment techniques addressing sensitivity and intrusion issues. *Ergonomics* 27: 1033–1050, 1984.

Derrick, W.A. The dissociation between subjective and performance-based measures of operator workload. Proceedings of the National Aerospace and Electronics Conference, 1020–1025, 1985.
Hancock, P.A. Effects of control order, augmented feedback, input device and practice on tracking performance and perceived workload. *Ergonomics* 39(9): 1146–1162, 1996.
King, T., Hamerman-Matsumoto, J., and Hart, S.G. Dissociation revisited: Workload and performance in a simulated flight task. Proceedings of the 5th International Symposium on Aviation Psychology, 796–801, 1989.
Kramer, A.F., Sirevaag, E.J., and Braune, R. A psychophysiological assessment of operator workload during simulated flight missions. *Human Factors* 29: 145–160, 1987.
Madero, R.P., Sexton, G.A., Gunning, D., and Moss, R. *Total Aircrew Workload Study for the AMST (AFFDL-TR-79-3080 Volume 1)*. Wright-Patterson Air Force Base, OH: Air Force Flight Dynamics Laboratory, February 1979.
Stein, E.S. *The Measurement of Pilot Performance; A Master-Journeyman Approach (DOT/FAA/CT-83/15)*. Atlantic City, NJ: Federal Aviation Administration Technical Center, May 1984.
Thornton, D.C. An investigation of the "Von Restorff" phenomenon. Proceedings of the Human Factors 29th Annual Meeting, 760–764, 1985.
Vidulich, M.A., and Bortolussi, M.R. A dissociation of objective and subjective workload measures in assessing the impact of speech controls in advanced helicopters. Proceedings of the 32nd Annual Meeting of the Human Factors Society, 1471–1475, 1988.
Vidulich, M.A., and Wickens, C.D. Causes of dissociation between subjective workload measures and performance: Caveats for the use of subjective assessments. Proceedings of the 3rd Symposium on Aviation Psychology, 223–230, 1985.
Vidulich, M.A., and Wickens, C.D. Causes of dissociation between subjective workload measures and performance. *Applied Ergonomics* 17(4): 291–296, 1986.
Wickens, C.D., and Liu, Y. Codes and modalities in multiple resources: A success and a qualification. *Human Factors* 30: 599–616, 1988.
Wickens, C.D., and Yeh, Y. The dissociation between subjective workload and performance: A multiple resource approach. Proceedings of the Human Factors Society 27th Annual Meeting, 244–248, 1983.
Wierwille, W.W., Rahimi, M., and Casali, J.G. Evaluation of 16 measures of mental workload using a simulated flight task emphasizing mediational activity. *Human Factors* 27: 489–502, 1985.
Wolf, J.D. *Crew Workload Assessment: Development of a Measure of Operator Workload (AFFLD-TR-78-165)*. Wright-Patterson Air Force Base, OH: Air Force Flight Dynamics Laboratory, December 1978.
Yeh, Y., and Wickens, C.D. The effect of varying task difficulty on subjective workload. Proceedings of the Human Factors Society 29th Annual Meeting, 765–769, 1985.
Yeh, Y., and Wickens, C.D. Dissociation of performance and subjective measures of workload. *Human Factors* 30: 111–120, 1988.

List of Acronyms

3D	Three Dimensional
a	number of alternatives per page
AET	Arbeitswissenschaftliches Erhebungsverfahren zur Tatigkeitsanalyze
AGARD	Advisory Group for Research and Development
AGL	Above Ground Level
AHP	Analytical Hierarchy Process
arcmin	arc minute
ATC	Air Traffic Control
ATWIT	Air Traffic Workload Input Technique
AWACS	Airborne Warning And Control System
BAL	Blood Alcohol Level
BVR	Beyond Visual Range
c	computer response time
C	Centigrade
CARS	Crew Awareness Rating Scale
CC-SART	Cognitive Compatibility Situational Awareness Rating Scale
cd	candela
CLSA	China Lake Situational Awareness
cm	centimeter
comm	communication
C-SWAT	Continuous Subjective Workload Assessment Technique
CTT	Critical Tracking Task
d	day
dBA	decibels (A scale)
dBC	decibels (C scale)
EAAP	European Association of Aviation Psychology
F	Fahrenheit
FOM	Figure of Merit
FOV	Field of View
ft	Feet
GCI	Ground Control Intercept
G_y	Gravity y axis
G_z	Gravity z axis
h	hour
HPT	Human Performance Theory
HSI	Horizontal Situation Indicator
HUD	Head Up Display
Hz	Hertz
i	task index

183

ILS	Instrument Landing System
IMC	Instrument Meteorological Conditions
in	inch
ISA	Instantaneous Self Assessment
ISI	Interstimulus interval
j	worker index
k	key press time
kg	kilogram
kmph	kilometers per hour
kn	knot
KSA	Knowledge, Skills, and Ability
LCD	Liquid Crystal Display
LED	Light Emitting Diode
LPS	Landing Performance Score
m	meter
m²	meter squared
mg	milligram
mi	mile
min	minute
mm	millimeter
mph	miles per hour
msec	milliseconds
MTPB	Multiple Task Performance Battery
nm	nautical mile
NPRU	Neuropsychiatric Research Unit
OW	Overall Workload
PETER	Performance Evaluation Tests for Environmental Research
POMS	Profile of Mood States
POSWAT	Pilot Objective/Subjective Workload Assessment Technique
PPI	Pilot Performance Index
ppm	parts per million
PSE	Pilot Subjective Evaluation
r	total number of index pages accessed in retrieving a given item
rmse	root mean squared error
RT	reaction time
RWR	Radar Warning Receiver
s	second
SA	Situational Awareness
SAGAT	Situational Awareness Global Assessment Technique
SALIENT	Situational Awareness Linked Instances Adapted to Novel Tasks
SART	Situational Awareness Rating Technique
SA-SWORD	Situational Awareness Subjective Workload Dominance
SD	standard deviation

List of Acronyms

SPARTANS	Simple Portable Aviation Relevant Test Battery System
st	search time
STOL	Short Take-Off and Landing
STRES	Standardized Tests for Research with Environmental Stressors
SWAT	Subjective Workload Assessment Technique
SWORD	Subjective WORkload Dominance
t	time required to read one alternative
TEWS	Tactical Electronic Warfare System
TLC	Time to Line Crossing
TLX	Task Load Index
t_z	integration time
UAV	Uninhabited Aerial Vehicle
UTCPAB	Unified Tri-services Cognitive Performance Assessment Battery
VCE	Vector Combination of Errors
VDT	Video Display Terminal
VMC	Visual Meteorological Conditions
VSD	Vertical Situation Display
WB	bottleneck worker
WCI/TE	Workload/Compensation/Interference/Technical Effectiveness

Author Index

Abbott, T.S. 114, 129, 136, 142
Abdelrahman, A. 119, 128
Acton, W. 38, 39, 136, 138, 139, 140, 141, 143, 180
Aiken, E.W. 85, 86
Albery, W. 135, 140
Allen, R.W. 17, 18
Allport, D.A. 26, 29, 31
Alluisi, E.A. 8, 9, 10, 29, 32, 41, 43, 45, 47, 48, 71
Amell, J.R. 13, 14
Ames, L.L. 97, 100
Anatasi, J.S. 60, 62
Anderson, P.A. 40
Andre, A.D. 33, 35, 53, 54, 67, 69
Antonis, B. 26, 27, 29, 31
Ashby, M.C. 7, 8

Bahrick, H.P. 34, 35
Ball, J. 58, 61, 99, 100, 134, 141
Bateman, R. 92, 136
Bateman, R.P. 37, 39, 140, 141, 142, 160, 180
Baty, D.L. 17, 20
Beare, A. 136, 140
Beatty, J. 34, 36
Becker, A.B. 117, 123
Becker, C.A. 17, 18, 53, 54
Beer, M.A. 20, 21
Bell, P.A. 41, 43
Benson, A.J. 17, 18
Bergeron, H.P. 41, 43, 50
Bierbaum, C.R. 161, 162, 163
Biers, D.W. 121, 129, 136, 137, 139, 140, 142, 154, 156
Birkmire, D.P. 60, 62
Bittner, A.C. 67, 69, 70, 80, 88, 91, 92, 110, 112, 115, 121, 123, 124, 127, 138, 142, 147, 154, 156
Bivens, C.C. 107, 109, 135, 136, 138, 142
Bloomberg, R. 149, 150
Bobko, D.J. 64, 66
Bobko, P. 64, 66

Boff, K.R. 5, 6, 79, 89, 147
Boggs, D.H. 41, 43
Boies, S.J. 29, 32
Bonto, M.A. 60, 62
Borg, C.G. 78
Borresen, R. 91, 92, 136, 142
Bortolussi, M.R. 18, 19, 40, 43, 63, 65, 66, 76, 77, 78, 83, 84, 107, 109, 110, 112, 113, 120, 123, 131, 132, 137, 140, 151, 169, 181
Boyce, P.R. 68, 69
Boyd, S.P. 137, 140
Braun, J.R. 133, 134
Braune, R. 11, 12, 78, 79, 168, 181
Bressler, J.R. 135, 143
Briggs, G.E. 60, 61
Broadbent, D.E. 29, 31
Brockman, W. 68, 70
Brouwer, W.H. 24, 25
Brown, I.D. 8, 14, 17, 18, 29, 31, 34, 35, 38, 39, 41, 43, 127
Brown, J.L. 43
Buck, J. 163
Budescu, D.V. 76, 77
Burdsal, C. 91, 92, 136, 142
Burrows, A.A. 41, 43
Butler, K. 29, 32, 41, 43
Byars, G.E. 97, 100
Byers, J.C. 91, 92, 112, 115, 121, 123, 124, 127, 138, 142, 154, 156

Carpenter, S. 21
Carsten, O. 47, 48
Casali, J.G. 90, 91, 92, 93, 104, 146, 147, 168, 169, 180, 181
Cashion, P.A. 33, 35, 53, 54, 67, 69
Cassenti, D.N. 136, 140
Chang, S.X. 120, 129
Chechile, R.A. 29, 32, 41, 43
Chesney, G.L. 17, 19
Childress, M.E. 64, 66
Cho, D.W. 133, 134
Chow, S.L. 29, 32, 41, 43

187

Christ, R.E. 83, 84, 91, 92, 112, 115, 123, 124, 127, 138, 142, 154, 156, 161, 162
Cibelli, L.A. 133, 134
Clement, W.F. 67, 69
Colle, H. 13, 14, 91, 93, 136, 139, 140, 145
Comstock, E.M. 54
Connor, S. 34, 36, 60, 63, 65
Connor, S.A. 86, 91, 93, 146, 147
Conrad, R. 10, 11
Constantini, A.F. 133
Cook, M.R. 136, 141
Cooper, G.E. 85, 86
Cordes, R.E. 25, 26, 27, 52
Corkindale, K.G.G. 67, 69
Corlett, E.N. 6, 15, 75
Costantini, A.F. 134
Courtright, J.F. 99, 100, 136, 141
Crabtree, M.S. 38, 39, 85, 86, 136, 141, 180
Crawford, B.M. 60, 61
Crombie, R.B. 143
Cross, K.C. 161, 162
Cumming, F.G. 67, 69

Damos, D. 17
Daniel, J. 29, 32
Davies, A.K. 6
Davis, D. 162, 163
Davis, J.E. 133, 134
Davis, J.M. 58, 62
Davis, M.A. 64, 66
Deimler, J.D. 162, 163
Dellinger, J.A. 58, 61, 62
Derrick, W.L. 14, 137, 141, 166
Deschesne Burkhardt 120, 129
Dessouky, M.I. 151
Detweiler, M. 13, 15, 17, 18
Dick, A.O. 16, 19, 21, 23, 24, 25, 27, 28, 32, 36, 46, 48, 49, 50, 51, 55, 56, 57, 61, 63, 66, 70, 80, 88, 110, 147
Dietrich, C.W. 46
DiMarco, R.J. 17, 18
Donchin, E. 17, 19, 34, 35, 36, 41, 44, 56
Donnell, M.L. 88
Dorfman, P.W. 41, 43
Dornic, S. 41, 43
Dorris, R. 136, 140
Drum, J.E. 115, 129

Dunbar, S. 17, 19, 54, 55
Dunleavy, A.O. 100, 101
Dunn, R.S. 58, 62

Eberts, R. 68, 69, 70
Eggemeier, F.T. 3, 5, 6, 74, 75, 78, 79, 89, 113, 117, 121, 123, 129, 135, 136, 139, 141, 142, 143, 146, 147
Eggleston, R.G. 154, 156
Ellis, J.E. 17, 18, 41, 43
Ellis, K. 112, 127
Endsley, M.R. 118, 125, 127
Erdos, R 127
Erdos, R. 112
Ewry, M.E. 13, 14

Fadden, D. 79, 80
Fairclough, S.H. 7, 8
Farber, E. 46
Ferneau, E.W. 133, 134
Ferris, T.K. 117, 130
Figarola, T.R. 41, 43
Finegold, L.S. 100, 101
Finell, G. 59, 61, 67, 70
Finkelman, J.M. 29, 32, 33, 51
Fisher, R.P. 60, 61, 68, 70
Fisher, S. 17, 18, 34, 35, 133, 134
Fistel, A.L. 117, 125
Fitts, P.M. 34, 35
Fleishman, E.A. 41, 43
Florek, H. 29, 32
Flux, R. 33, 34, 36
Fort, A. 149, 150
Fouillot, J. 149, 150
Fournier, B.A. 26, 27
Fowler, F. 29, 32
Fox, J.G. 15
Fracker, M.L. 136, 141
Frankenfeld, C.A. 99, 100
Frazier, M.L. 143
Freeman, M.H. 36
Fulford, L.A. 162
Fuller, N. 6
Funke, G. 119, 129

Gabay, E. 26, 27
Gallagher, V. 46
Gallaway, R.A. 20, 21
Garinther, G. 136, 145

Author Index

Gartner, W.B. 3, 5
Garvey, W.D. 10, 11
Gawron, V.J. 12, 13, 16, 17, 18, 58, 59, 61, 68, 69, 99, 100, 133, 134, 137, 141, 167
George, E.J. 97, 99, 100, 101, 102
Gidcomb, C. 136, 139, 141
Gill, R.T. 136, 140
Giroud, Y. 17, 18
Glanzer, M. 34, 36
Glass, D.C. 29, 32
Gliva-McConvey 119, 129
Godthelp, H. 93, 94, 95
Goldstein, A.P. 133, 134
Goldstein, I.L. 41, 43
Goodyear, C. 135, 140
Graham, C.H. 43, 136, 141
Green, R. 33, 34, 36
Greenberg, S.N. 22, 23, 29, 32, 68, 70
Gregory, M. 29, 31
Griffiths, I.D. 68, 69
Groth, H. 47, 48
Guadagnoli, E. 133, 134
Gunning, D. 64, 65, 66, 168, 181
Gutmann, J.C. 26, 27, 52
Gutowski, W. 29, 32, 41, 43

Halcomb, C.G. 41, 44
Hall, B.J. 122, 128
Hallbeck, S. 119, 128, 133
Hamilton, D.B. 161, 162, 163
Hamilton, K.M. 121, 126
Hammerton-Fraser, A.M. 67, 69
Hancock, P.A. 117, 118, 123, 124, 126, 127, 132, 133, 134, 136, 138, 141, 142, 143, 154, 156, 168, 181
Hann, R.L. 136, 144
Hansen, M.D. 68, 69
Harbeson, M.M. 67, 69
Harms, L. 35, 36
Harper, R.P. 85, 86
Harris, R.M. 83, 84, 92, 156, 161, 162
Harris, W.C. 118, 126, 134, 154, 156
Hart, S.G. 17, 18, 19, 40, 43, 60, 61, 63, 64, 65, 66, 78, 102, 103, 107, 109, 110, 111, 112, 113, 122, 126, 131, 137, 142, 151, 180, 181
Hartman, B.O. 5, 61
Hartzell, E.J. 64, 66
Haskell, B.E. 137, 139, 142

Haskell, D.H. 133, 134, 135
Haskell, G. 21
Hassoun, J.A. 136, 138, 144, 145
Hauser, J.R. 64, 66, 102, 103, 112, 126
Haworth, L.A. 107, 109, 135, 136, 138, 142
Hayes, C. 116, 125
Haygood, R.C. 16, 19
Heers, S.T. 33, 34, 35, 53, 54, 67, 69, 114, 126
Helm, W. 78
Hendy, K.C. 121, 126
Her, C. 159
Herman, L.M. 41, 43
Heron, A. 29, 31
Hess, R.A. 68, 70
Hicks, M.R. 6
Hilgendorf, E.L. 41, 43
Hill, S.G. 16, 19, 21, 23, 24, 25, 27, 28, 32, 36, 39, 45, 46, 48, 49, 50, 51, 55, 56, 57, 61, 63, 66, 70, 80, 83, 84, 88, 91, 92, 110, 112, 115, 121, 123, 124, 127, 138, 142, 147, 154, 156, 161, 162
Hoffman, E.R. 43
Hoffman, M. 60, 61
Hohmuth, A.V. 41, 43
Hollis, S. 99, 100
Holloway, C.M. 41, 44, 68, 71
Huddleston, J.H.F. 17, 18, 29, 32, 33, 34, 36, 41, 44
Hughes, E. 138, 145
Hwang, S. 159
Hyman, F. 59, 62

Iavecchia, H.P. 91, 92, 115, 127, 138, 142, 154, 156, 162, 163
Iervolino, A. 133, 134
Israel, J.B. 17, 19

Jackson, M. 120, 129
Jamson, A.H. 47, 48
Jaschinski, W. 34, 36
Jenkins, M.L. 13, 14
Jennings, A.E. 34, 35, 41, 43
Jensen, R.S. 13
Jensen, S.E. 149
Jex, H.R. 17, 18
Johannsen, G. 37, 39
Johnson, D.F. 16, 19

Author Index

Johnson, J. 100, 101
Johnson, W.W. 114, 120, 132, 148, 149
Johnston, N. 6
Johnston, W.A. 68, 70
Jones, B.W. 41, 44
Jones, E.L. 120, 129
Jorbert, P.N. 41, 43
Jordan, P.W. 114, 127

Kahneman, D. 34, 36
Kalsbeek, J.W.H. 34, 36, 47, 48, 49, 50
Kantowitz, B.H. 17, 18, 19, 32, 34, 36, 38, 39, 40, 43, 63, 65, 66, 107, 109
Kantowitz, S.C. 17, 19, 107, 110
Kappler, W.D. 93, 94, 95
Kaufman, L. 5, 6, 75, 79, 89, 147
Keele, S.W. 29, 32, 34, 36
Keillor, J. 112, 127
Kelley, C.R. 41, 44
Kelley, T.D. 136, 140
Kelly, C.R. 22
Kelly, P.A. 16, 17, 19, 53, 54, 55
Kelly, R.T. 54, 55
Kenny, C.A. 115, 125
Keppel, G. 1
Kessel, C. 29, 33, 68, 70
Kijowski, B.A. 151
Kilmer, K.J. 91, 92, 136, 142
King, A.H. 76, 77
Kirk, R.R. 1
Klapp, S.T. 16, 17, 19, 53, 54, 55
Klaus, J. 159, 160
Klein, G.A. 29, 32
Kleinbeck, U. 68, 70
Knapp, B.G. 122, 128
Knapp, R. 91, 92, 136, 142
Knight, J.L. 34, 36, 38, 39
Knotts, L.H. 58, 59, 61
Knowles, W.B. 10, 11, 14, 15, 40, 44
Kobus, D.A. 21
Koritsas, E. 114, 130
Kosinar, V. 29, 32
Kramer, L.J. 99, 100
Krebs, M.J. 85, 86
Kristofferson, A.B. 46
Krol, J.P. 17, 19
Kuperman, G. 136, 137, 141, 142
Kyriakides, K. 41, 44

Labacqz, J.V. 85
Lai, F.C.H. 47, 48
Landry, L.N. 121, 126
Langolf, G.D 58, 62
LaPointe, P. 136, 141
Laudeman, I.V. 9
Laurell, H. 54, 55
Laurencelle, L. 17, 18
Lavalliere, M. 29, 32
Lawless, M.T. 100, 101
Lay, J.E. 136, 144
Lebacqz, J.V. 86
Lee, S.W. 28, 32, 67, 70
LeGrand, J.E. 58, 62, 128
Leopold, F.F. 34, 36, 47, 48, 49, 50
Lester, P.T. 107, 109
Leventhal, H.G. 41, 44
Levison, W.H. 46
Lidderdale, I.G. 76, 77, 83, 84
Lindsay, P.H. 41, 44
Linton, P.M. 16, 19, 21, 23, 24, 25, 27, 28, 32, 36, 39, 45, 46, 48, 49, 50, 51, 55, 56, 57, 58, 60, 61, 62, 63, 66, 70, 80, 88, 110, 147, 154, 156
Lisper, H.L. 54, 55
Liu, Y.Y. 65, 66
Lloyd, M. 58, 61, 99, 100, 134, 141
Logan, G.D. 17, 19
Long, J. 41, 44
Looper, M. 68, 70
Lorenz, B. 59, 61, 67, 70
Lorr, M. 133, 134
Lowenstein, L.M. 133, 134
Lozano, M.L. 101, 102
Lu, L.G. 80
Lucaccini, L.F. 47, 48
Lundy, D.H. 17, 18
Lutmer, P.A. 136, 142
Lyman, J. 47, 48, 164
Lysaght, R.J. 15, 16, 17, 19, 20, 21, 22, 23, 24, 25, 26, 27, 28, 29, 31, 32, 34, 35, 36, 37, 38, 39, 40, 41, 45, 46, 47, 48, 49, 50, 51, 53, 54, 55, 56, 57, 59, 60, 61, 63, 64, 65, 66, 67, 68, 69, 70, 79, 80, 83, 84, 88, 92, 108, 110, 147, 154, 156, 161, 162

MacKinnon, D.P. 163, 164
Madero, R.P. 64, 66, 168, 181

Author Index

Malmstrom, F.V. 41, 44
Malzahn, D. 91, 92, 136, 142
Mandler, G. 29, 32
Manning, C.M. 126, 136, 142
Manning, G. 113
Manzey, D. 67, 70, 116, 120, 128
Marks, L.B. 120, 129
Martin, D.W. 54, 55, 68, 70
Masline, P.J. 151
Matsumoto, J.H. 112, 131
Mazur, L.M. 120, 129
McCracken, J.H. 160, 161
McDonald, N. 6
McDonnell, J.D. 154, 155
McGhee, J.Z. 139, 141
McGlothlin, W. 29, 32
McGrath, J.J. 41, 44
McGregor, E.A. 136, 140
McKenzie, R.E. 5
McLeod, P.D. 34, 36
McNair, D.M. 133, 134, 135
McRuer, D.T. 17, 18
Meador, M. 59, 62
Mehtaki, N. 143
Meister, D. 1
Melville, B. 136, 141
Merat, N. 47, 48
Merhav, S.J. 26, 27
Meshkati, N. 5, 6, 14, 15, 75, 78, 126
Metta, D.A. 76, 77
Meyer, R.E. 133, 134
Micalizzi, J. 59, 61
Michaux, W.W. 133, 134
Michon, J.A. 38, 39
Miller, J. 58, 61, 99, 100, 134, 141
Miller, K. 20, 21, 54, 55
Miller, R.C. 12, 13, 60, 61, 97, 100, 107, 110
Milone, F. 29, 33
Mirchandani, P.B. 68, 70
Mirin, S.M. 133, 134
Mitchell, J.A. 121, 129
Mitsuda, M. 26, 27, 41, 44
Miyake, S. 113, 126, 136, 142
Monty, R.A. 41, 44
Moray, N. 5, 39, 151
Moroney, W.F. 121, 129, 136, 142
Mosaly, P.R. 120, 129
Moskowitz, H. 29, 32

Moss, R. 64, 66, 168, 181
Mountford, S.J. 22, 23, 29, 33, 34, 36, 68, 70
Muckler, F.A. 74, 75
Murdock, B.B. 15, 16, 29, 32, 41, 43
Murphy, M.R. 3, 5
Muter, P. 41, 44
Muto, W.H. 26, 27, 52

Narayanan, S. 115, 130
Narvaez, A. 97, 100
Nathan, P.F. 133, 134
Netick, A. 16, 17, 19
Newell, F.D. 85, 86
Newlin, E.P. 10, 11
Newlin-Canzone 119, 129
Noble, M. 29, 32, 34, 35, 47, 48, 51
Norcross, J.C. 133, 134
Nordeen, M. 99, 100
Norman, D.A. 41, 44
North, R.A. 159, 160
Notestine, J. 136, 142
Ntuen, C.A. 86, 163, 164
Nygren, T.E. 113, 129, 135, 136, 138, 143

O'Connor, M.F. 88, 89
O'Donnell, R.D. 3, 5, 6, 60, 61, 67, 70, 74, 75, 78, 79, 89, 146, 147

Pak, R. 115, 129
Palmer, E.A. 9, 29, 32, 41, 43
Panganiban, A.R. 119, 129
Papa, R.M. 91, 92, 136, 143
Park, E. 86
Park, K.S. 28, 32, 67, 70
Parke, R.C. 97, 99, 100, 133, 135
Parker, F. 58, 61, 99, 100, 134, 141
Parkes, A.M. 7, 8
Patterson, J.F. 88
Pearson, R.G. 97, 100
Pearson, W.H. 60, 61
Pepitone, D.D. 112, 131
Peters, G.L. 60, 61
Peters, L. 136, 145
Peters, L.J. 60, 62
Pfendler, C. 37, 39, 93, 94, 95
Pillard, R.C. 133, 134
Pinkus, A.R. 67, 70
Pitrella, F.D. 93, 94, 95

Plamondon, B.D. 16, 19, 21, 23, 24, 25, 27, 28, 32, 36, 39, 45, 46, 48, 49, 50, 51, 55, 56, 57, 61, 63, 66, 70, 80, 88, 110, 147
Plath, G. 113, 125
Pollack, I. 34, 36
Pollack, J. 137, 143
Pollock, V. 133, 134
Porges, S.W. 58, 61, 62
Porterfield, D.H. 7, 157, 158
Poston, A.M. 58, 62
Potter, S.S. 135, 136, 143
Poulton, E.C. 13, 15, 29, 31, 34, 35
Previc, R.H. 20, 21
Price, D.L. 26, 27
Prochaska, J.O. 133, 134
Proteau, L. 17, 18
Pugatch, D. 133, 134, 135
Purvis, B. 136, 144
Putz, V.R. 41, 44

Quinn, T.J. 154, 156

Rahimi, M. 146, 147, 169, 181
Ramacci, C.A. 33, 36, 67, 70
Randle, R.J. 41, 44
Rapoport, A. 76, 77
Raskin, A. 133, 134
Raza, H. 114, 132
Reed, L.E. 41, 44
Rehman, J.T. 158
Reid, G.B. 135, 136, 137, 138, 139, 140, 141, 142, 143, 144, 145
Reiger, C. 93
Reker, D. 133, 134
Repperger, D.W. 80
Reynolds, P. 26, 27, 29, 31
Richardson, B.C. 58, 62
Rieger, C.A. 91, 93
Robinson, C.P. 69, 70
Rockwell, J. 120, 129
Rodriguez Paras, C. 117, 130
Roe, M. 135, 140
Roediger, H.L. 34, 36, 38, 39
Rokicki, S.M. 99, 100, 138, 140
Rolfe, J.M. 14, 15, 17, 18
Romer, W. 93

Roner, W. 39, 67, 147
Roscoe, A.H. 81, 83, 84, 144, 150
Roscoe, S.N. 13, 15
Rosenberg, B.L. 157, 158
Rossi, A.M. 133, 134
Rota, P. 33, 36, 67, 70
Roth, I. 133, 134
Rothe, R. 41, 44
Rothengatten, T. 24, 25
Rozovski, D. 112, 127
Ruby, W.J. 41, 44
Rueb, J. 138, 144
Ruff 80, 115, 130
Ruggerio, F. 79, 80
Russotti, J. 21

Saaty, T.L. 76, 77
Savage, R.E. 25, 27, 52
Schaffer, A 46, 47, 48
Schick, F.V. 136, 144
Schiewe, A. 59, 61, 67, 70
Schiflett, S.G. 12, 13, 57, 58, 59, 60, 62, 99, 100, 134, 141
Schlichting, C. 21
Schmidt, K.H. 68, 70
Schneider, W. 13, 14, 15
Schohan, B. 41, 44
Schori, T.R. 41, 44
Schouten, J.F. 34, 36, 47, 48, 49, 50
Schreiner, W. 22, 23, 29, 33, 34, 36, 68, 70
Schueren, J. 80, 81
Schultz, W.C. 85, 86
Schvaneveldt, R.W. 17, 19
See, J.E. 137, 144, 163, 164
Seidelman 64, 67, 116, 120, 130, 131
Seidner, J. 142
Senders, J.W. 46
Seven, S.A. 74, 75
Sexton, G.A. 64, 66, 168, 181
Shachem, A. 133, 135
Shapiro, L.M. 133, 134
Shelton, K.J. 99, 100
Shingledecker, C.A. 38, 39, 135, 136, 141, 143
Shively, R.J. 17, 18, 19, 40, 43, 60, 61, 63, 65, 66, 107, 109, 110, 112, 115, 125, 131, 135, 136, 138, 142

Author Index

Shulman, H.G. 23, 29, 32
Silverstein, C. 34, 36
Simmonds, D.C.V. 38, 39
Simons, J.C. 136, 137, 140, 144
Simons, J.L. 100, 101
Singleton, J.G. 15
Sirevaag, E.J. 78, 79, 168, 181
Skelly, J.J. 136, 144
Skipper, J.H. 91, 93
Slater, T. 58, 61, 99, 100, 134, 141
Slocum, G.K. 13, 15
Smith, M.C. 17, 20
Smith, P.J. 58, 62
Smith, R.L. 47, 48
Soliday, S.M. 41
Solomon, P. 133, 134
Speyer, J. 149, 150
Spicuzza, R.J. 58, 60, 61, 62, 67, 70, 99, 100, 134, 141
Stadler, M. 136, 141
Stager, P. 26, 27, 29, 33, 41, 44
Staveland, L.E. 78, 111, 126
Stein, E.S. 102, 148, 149, 157, 158, 168, 181
Stein, W. 37, 39
Stening, G. 54, 55
Sternberg, S. 57, 62
Stokes, J. 162, 163
Stoliker, J.R. 91, 92, 136, 143
Storm, W.F. 97, 99, 100, 133, 135
Strickland, D. 86
Strizenec, M. 29, 32
Swink, J. 47, 48
Szabo, S.M. 161, 162

Taylor, H.L. 58, 60, 61, 62
Taylor, R.M. 114, 130
Teichgraber, W.M. 68, 70
Thiele, G. 59, 61, 67, 70
Thiessen, M.S. 136, 144
Thomas, J.P. 5, 6, 75, 89, 147
Thompson, M.W. 160
Thurmond, D. 99, 100
Tickner, A.H. 17, 18, 38, 39
Titler, N.A. 133, 134
Toivanen, M.L. 40
Tomoszek, A. 6

Travale, D. 58, 61, 99, 100, 134, 141
Truijens, C.L. 51
Trumbo, D. 29, 32, 33, 47, 48, 51
Tsang, P.S. 29, 33, 56, 57, 59, 60, 62, 68, 70, 76, 78, 80, 83, 84, 107, 110, 114, 120, 132, 136, 138, 144, 156
Tyler, D.M. 41, 44

van de Graaff, R.C. 154
van Merrienboer, J.J.G 10
van Wolffelaar, P.C. 24, 25
Velaquez, V.L. 60, 62
Vickroy, C.C. 144, 160
Vickroy, S.C. 136
Vidulich, M.A. 13, 15, 56, 57, 59, 60, 61, 62, 76, 77, 78, 80, 81, 83, 84, 107, 110, 113, 118, 119, 120, 124, 132, 136, 137, 138, 144, 156, 163, 164, 169, 181
Volavka, J. 133, 134
Vroon, P.A. 38, 39
Vu, K.P.L. 132, 148, 149

Wagenaar, W.A. 51
Wainwright, W. 81, 84
Wallace, A. 119, 129
Ward, G.F. 81, 136, 144
Ward, J.L. 46
Ward, S.L. 136, 140
Wargo, M.J. 21, 22, 41, 44
Warr, D. 91, 93, 136, 145
Waterink, W. 24, 25
Watson, A.R. 86, 163, 164
Weinstein, G.J. 133, 134
Weller, M.H. 58, 62
Wempe, T.E. 17, 20
Wenger, M.J. 60, 62
Wetherell, A. 14, 15, 29, 33, 34, 36, 41, 44, 47, 48, 51, 60, 62
Wherry, R.J. 16, 19, 21, 23, 24, 25, 27, 28, 32, 36, 39, 45, 46, 48, 49, 50, 51, 55, 56, 57, 61, 63, 66, 70, 80, 88, 110, 147
Whitaker, L.A. 68, 70, 136, 145
Whitbeck, R.F. 85, 86
White, J. 6
White, S.A. 163, 164

Whitfield, D. 15
Wickens, C.D. 11, 12, 14, 15, 17, 19, 22, 23, 29, 33, 34, 35, 36, 41, 44, 50, 54, 55, 56, 58, 59, 60, 61, 62, 65, 66, 68, 70, 74, 75, 164, 165, 166, 169, 178, 181
Wiegand, D. 93, 94, 95
Wierwille, W.W. 3, 5, 6, 7, 8, 13, 15, 16, 17, 19, 21, 23, 24, 25, 26, 27, 28, 31, 32, 33, 34, 36, 37, 38, 39, 45, 46, 48, 49, 50, 51, 52, 55, 56, 57, 60, 61, 63, 64, 65, 66, 67, 70, 74, 75, 80, 85, 86, 88, 89, 90, 91, 92, 93, 104, 110, 146, 147, 150, 151, 168, 169, 180,181
Williams, G. 113, 126, 136, 142
Williges, R.C. 15, 67, 70
Wilson, G.F. 138, 145
Wilson, G.R. 136, 144
Wilson, J.R. 6, 15, 75
Wilson, R.V. 29, 32, 33, 34, 36, 41, 44

Wingert, J.W. 85, 86
Wojtowicz, J. 21
Wolf, J.D. 57
Worden, P.E. 29, 32
Wright, P. 41, 44, 68, 71

Xu, J. 120, 129

Yeh, Y. 11, 12

Zachary, W. 162, 163
Zaklad, A.L. 16, 19, 21, 23, 24, 25, 27, 28, 32, 36, 39, 44, 45, 46, 48, 49, 50, 51, 55, 56, 57, 61, 63, 66, 70, 74, 80, 88, 91, 92, 110, 112, 115, 123, 124, 127, 138, 142, 147, 154, 156, 162, 163
Zare, N.C. 133, 134
Zeitlin, L.R. 7, 24, 29, 33, 51
Zingg, J.J. 136, 141
Zufelt, K. 29, 33
Zwick, R. 76, 77

Subject Index

absolute error 64
acceleration 24, 86, 140
accuracy 39, 58, 59
AET 159
AHP 75, 76, 77, 156, 178, 179
AIM-s 96
Air Traffic Control 95, 151, 158
Air Traffic Workload Input Technique 7, 148, 157
aircraft simulator 6, 112, 113, 114, 120, 136, 137, 154, 167
Aircrew Workload Assessment System 6
airdrop 99, 167
altitude 44, 55, 59, 144
Analytical Hierarchy Process 72, 75, 121, 156
approach 99, 137, 157
Arbeitswissenshaftliches Erhebungsverfahren zur Tatigkeitsanalyze 159
Assessing the Impact of Automation on Mental Workload 95
attention 13, 14, 18, 19, 22, 25, 27, 31, 32, 39, 43, 44, 46, 48, 50, 53, 55, 61, 63, 70, 76, 135
automation 40, 66, 76, 113, 118

Bedford Workload Scale 72, 81, 82, 83, 121, 138, 149, 152, 153

California Q-Sort 119, 138
card sorting 15, 51
centrifuge 167
choice RT 16, 31, 35, 37, 40, 53, 69, 70
classification 20, 22, 29, 31, 53, 54, 68, 69, 160
Cognitive Failures Questionnaire 119, 138
Cognitive Interference Questionnaire 119, 138
combat 167
command and control 136
communication 83, 90

comparison measures 71
comprehensive 14, 16, 19, 21, 23, 24, 25, 27, 28, 32, 36, 39, 44, 45, 46, 48, 49, 50, 51, 55, 56, 57, 61, 63, 66, 70, 80, 88, 110, 162
Computed Air Release Point 64
Computerized Rapid Analysis of Workload 72, 159, 160
Continuous Subjective Assessment of Workload 148, 149
control movements/unit time 6
Cooper-Harper Rating Scale 72, 73, 81, 84, 85, 89, 91, 136, 157
correlation 6, 7, 9, 11, 78, 107, 120, 121, 136, 137, 138, 162, 163
CRAWL 160
Crew Status Survey 72, 95, 97, 99
cross-adaptive loading secondary task 21
C-SAW 149
C-SWAT 156

DALI 121
decision tree 71, 81, 84
Defense Technical Information Center 167
detection 22, 25, 26, 29, 31, 34, 35, 37, 38, 40, 41, 53, 54, 68, 69
detection time 45
deviations 16, 158
diagnosticity 3, 94
dichotic listening 11
dissociation 3, 164, 165, 166, 167, 178
distraction secondary task 23
driving 16, 17, 24, 25, 26, 29, 31, 34, 35, 37, 38, 40, 41, 45, 46, 47, 51, 53, 54, 59, 60
Driving 50
driving secondary task 24
dual task 18, 24, 35, 59, 69, 140
Dual task 138
dynamic workload scale 149
Dynamic Workload Scale 72, 148, 149, 150

195

Subject Index

ease of learning 14
emergency room 120
equal-appearing intervals 72, 148, 150
error 70
Eurocontrol recording and Graphical display On-line 152
experimental method 1

fatigue 31, 32, 36, 43, 53, 67, 97, 99, 100, 102, 107, 109, 126, 133, 134, 156
feedback 123, 126
Finegold Workload Rating Scale 72, 95, 100
flight simulation 16, 17, 37, 38, 40, 41, 64, 65, 114
flight task 35, 39, 53, 58, 67, 75, 81, 92, 93, 104, 113, 114, 144, 146, 147
Flight Workload Questionnaire 72, 95, 102

glance 7
glance duration 7, 8
glance frequency 8

Hart and Bortolussi Rating Scale 72, 148, 151
Hart and Hauser Rating Scale 73, 95, 102, 103
Head Up Display 80, 112
Hick's Law 9
Honeywell Cooper-Harper Rating Scale 87
Honeywell Copper-Harper Rating Scale 81
hover 3
human performance 1, 5, 6, 32, 39, 75, 79, 89, 147
Human Robot Interaction Workload Measurement Tool 103
identification 25, 26, 27, 29, 31, 40, 41, 95
identification/shadowing secondary task 25
Initial Point 64

Instantaneous Self Assessment 73, 148, 151, 152
intercept 57, 58, 161
interstimulus interval 59
interval production 37
ISA 151, 152, 153

Jenkins Activity Survey 119, 138

Keillor 127

laboratory 11, 35, 107, 136, 167
landing 55, 121, 151
lexical 27
lexical decision task 17, 53, 54
Line Oriented Flight Training 121
load stress 8
localizer 55
Low Altitude Navigation and Targeting Infrared System for Night 91

Magnitude Estimation 73, 75, 78
mathematical processing 74, 146
McCracken-Aldrich Technique 73, 159, 160
McDonnell Rating Scale 73, 148, 154, 155
memory 17, 22, 26, 28, 29, 31, 32, 34, 36, 37, 38, 39, 40, 41, 43, 47, 51, 53, 54, 57, 59, 68, 69, 136
memory search 14, 58, 62
memory-recall secondary task 28
memory-scanning secondary task 28
mental effort 90
mental effort load 135
mental math 29, 31, 35, 37, 38, 40, 41, 45
mental math task 33
mental workload 114
Michon Interval Production 37
Mission Operability Assessment Technique 73, 81, 88, 89
Modified Cooper-Harper 136, 154
Modified Cooper-Harper Rating Scale 81, 89, 91, 170, 171, 176, 178, 179
Modified Cooper-Harper Workload Scale 91
Modified Petri Nets 163
monitoring 16, 17, 25, 26, 29, 31, 34, 35, 37, 38, 39, 40, 41, 44, 45, 46, 47, 64, 65, 68, 69, 136
mood 74, 133
motivation 164
Multi-descriptor Scale 170, 178, 179
Multi-Descriptor Scale 73, 95, 104
Multidimensional Rating Scale 73, 95, 104
Multiple Attribute Task Battery 152

Subject Index

Multiple Resources Questionnaire 95, 106
Multiple Task Performance Battery 45, 71
Myers-Briggs Type Indicator 119, 138

NASA Bipolar Rating Scale 73, 95, 107, 108
NASA TLX 73, 82, 91, 95, 106, 110, 111, 112, 113, 114, 115, 117, 118, 119, 121, 129, 138, 152, 153, 156, 172, 178
NASA-TLX 180
Null Operation System Simulation 163
number of errors 8, 10, 21, 25, 40, 63

Observed Workload Area 9
occlusion secondary task 45
omission 25, 40
Overall Workload Scale 73, 91, 136, 138, 148, 154
overload 5, 110, 132, 164

participants 6, 8, 11, 12, 13, 16, 35, 37, 45, 50, 56, 75, 76, 78, 92, 93, 99, 102, 111, 115, 122, 133, 139, 146, 149, 151, 154
pattern discrimination 45
percent correct 10, 16
percent errors 58
percentage correct scores 20, 35
peripheral version display 121
Peripheral Vision Display 58
Pilot Objective/Subjective Workload Assessment Technique 7, 73, 157
Pilot Subjective Evaluation 74, 75, 79
pitch 7, 25, 31, 35, 53, 67, 86, 144
POMS 133, 134
POSWAT 7, 157, 158, 174, 178
Prediction of Performance 120
primary task 5, 13, 14, 16, 22, 23, 25, 31, 33, 35, 37, 40, 49, 51, 52, 53, 56, 59, 64, 66, 69
probe stimulus 28, 31
problem solving 17, 29, 37, 38, 40, 41, 45, 48, 68
problem-solving 69
production/handwriting secondary task 48

productivity 142
Profile of Mood States 74, 95, 133
Psychological Screening Inventory 133
psychological stress load 135
psychomotor 38, 49, 50
pyridostigmine bromide 58

quantitative 12

random digit generation 14
randomization secondary task 50
rate of gain of information 9
reaction time 9, 10, 16, 17, 18, 19, 20, 21, 23, 27, 28, 29, 31, 34, 35, 36, 37, 38, 40, 41, 47, 48, 49, 50, 53, 54, 55, 56, 57, 58, 59, 60, 63, 68, 69, 137
reading secondary task 52
relative condition efficiency 10
reliability 20, 31, 58, 76, 80, 82, 91, 114, 120, 136, 154
representativeness 14
resource competition 11, 138
resource-limited tasks 164
Revised Crew Status Survey 99
Rotter's Locus of Control 119, 138

SAINT 163
SART 130
secondary task 13, 14, 15, 16, 18, 22, 23, 24, 25, 27, 31, 32, 33, 35, 36, 37, 40, 45, 48, 49, 51, 52, 53, 56, 62, 64, 66, 75, 92
Sequential Judgment Scale 81, 93, 94
Short Subjective Instrument 121
short-term memory 14, 59
shrink rate 117, 136
simple RT 35, 37, 53, 69
simulated flight 33, 34, 37, 39, 53, 54, 55, 59, 60, 66, 67, 68, 75, 79, 90, 92, 93, 104, 113, 129, 136, 137, 146, 147
simulator 7, 9, 16, 18, 19, 23, 24, 25, 27, 31, 33, 36, 37, 40, 46, 58, 63, 64, 66, 67, 78, 82, 86, 107, 110, 121, 136, 140, 150, 151, 158
Situational Awareness 1
slope 21, 28, 31, 55, 57, 58, 161
spatial transformation 25, 56
speed 7, 10, 11, 24, 44, 57, 58, 62, 119

speed stress 10
speed-maintenance secondary task 57
S-SWAT 138
steering reversals 16, 24
Sternberg 11, 37, 38, 57, 58, 59, 62, 67, 122, 137, 138, 169
subjective measure 1, 11, 12, 14, 71, 74, 78, 143
Subjective Workload Assessment Technique 74, 95, 135, 160
Subjective Workload Dominance 74, 75, 80, 81
Subjective Workload Rating 173
subtraction task 56
SWAT 78, 82, 91, 92, 107, 110, 120, 121, 132, 135, 136, 137, 138, 139, 140, 141, 142, 143, 144, 151, 154, 163, 170, 172, 173, 178, 179, 180
synthetic work battery 63

takeoff 151
tapping 34, 35, 39
target identification 45
task analyses 163
Task Analysis Workload 74, 159, 161
Task Difficulty Index 11
task load 78, 124, 126
task load 3
TAWL 156, 161, 162
Team Workload 4
Team Workload Questionnaire 95
temazepam 99, 133
Thought Occurrence Questionnaire 119, 138
three phase code transformation secondary task 63
time estimation 64, 66
time load 135
time margin 12
time on target 69
time-estimation secondary task 63
tracking 11, 16, 17, 18, 19, 20, 22, 24, 25, 26, 29, 31, 32, 34, 35, 36, 37, 38, 40, 41, 44, 45, 47, 48, 49, 50, 51, 53, 54, 55, 56, 59, 60, 67, 68, 69, 70, 80, 86, 91, 93, 118, 120, 126, 134, 136, 138, 158, 165, 167, 170, 178, 179
training 5, 12, 19, 33, 43, 55, 58, 59, 69, 85, 99, 107, 122, 160

underload 5, 164
Unmanned Aerial Vehicle 115
utilization 74, 148, 159

validity 76, 82, 91, 120, 135, 154
verbal reasoning 14
Vertical Take-Off and Landing 87
vigilance 39, 40, 43, 44, 48, 112, 117, 119
visual matrix rotation task 56

W/INDEX 163
WCI/TE 146
workload 1, 3, 5, 6, 7, 8, 9, 10, 11, 12, 13, 15, 16, 18, 19, 21, 23, 24, 25, 27, 28, 31, 32, 33, 35, 36, 37, 39, 40, 43, 44, 45, 46, 48, 49, 50, 51, 52, 53, 55, 56, 57, 58, 61, 62, 63, 64, 66, 67, 69, 70, 71, 72, 74, 75, 77, 78, 79, 80, 81, 82, 83, 84, 85, 86, 87, 88, 89, 91, 92, 93, 94, 97, 98, 99, 100, 101, 102, 103, 104, 107, 108, 109, 110, 111, 113, 115, 117, 120, 121, 122, 123, 124, 126, 127, 129, 131, 132, 134, 135, 136, 137, 138, 139, 140, 141, 142, 143, 144, 145, 146, 147, 149, 150, 151, 154, 156, 157, 158, 159, 160, 161, 162, 163, 164
Workload 166
Workload Assessment Keyboard 148, 157, 158
Workload Differential Model 163
Workload Index 163
Workload Profile 122, 138
Workload Rating 177
Workload Rating System 157
Workload scale secondary task 71
Workload/Compensation/Interference/ Technical Effectiveness 74, 95, 146
Workload/Compensation/Interference/ Technical Effectiveness Scale 172, 176

yaw deviation 16, 24

Zachary/Zaklad Cognitive Analysis 74, 159